The publishing house tredition has created the series **TREDITION CLASSICS**. It contains classical literature works from over two thousand years. Most of these titles have been out of print and off the bookstore shelves for decades.

The book series is intended to preserve the cultural legacy and to promote the timeless works of classical literature. As a reader of a **TREDITION CLASSICS** book, the reader supports the mission to save many of the amazing works of world literature from oblivion.

The symbol of **TREDITION CLASSICS** is Johannes Gutenberg (1400 – 1468), the inventor of movable type printing.

With the series, tredition intends to make thousands of international literature classics available in printed format again – worldwide.

All books are available at book retailers worldwide in paperback and in hardcover. For more information please visit: www.tredition.com

tredition was established in 2006 by Sandra Latusseck and Soenke Schulz. Based in Hamburg, Germany, tredition offers publishing solutions to authors and publishing houses, combined with worldwide distribution of printed and digital book content. tredition is uniquely positioned to enable authors and publishing houses to create books on their own terms and without conventional manufacturing risks.

For more information please visit: www.tredition.com

Nitro-Explosives: A Practical Treatise

P. Gerald (Percy Gerald) Sanford

Imprint

This book is part of the TREDITION CLASSICS series.

Author: P. Gerald (Percy Gerald) Sanford
Cover design: toepferschumann, Berlin (Germany)

Publisher: tredition GmbH, Hamburg (Germany)
ISBN: 978-3-8491-5457-8

www.tredition.com
www.tredition.de

Copyright:
The content of this book is sourced from the public domain.

The intention of the TREDITION CLASSICS series is to make world literature in the public domain available in printed format. Literary enthusiasts and organizations worldwide have scanned and digitally edited the original texts. tredition has subsequently formatted and redesigned the content into a modern reading layout. Therefore, we cannot guarantee the exact reproduction of the original format of a particular historic edition. Please also note that no modifications have been made to the spelling, therefore it may differ from the orthography used today.

PREFACE.

In compiling the following treatise, my aim has been to give a brief but thoroughly practical account of the properties, manufacture, and methods of analysis of the various nitro-explosives now so largely used for mining and blasting purposes and as propulsive agents; and it is believed that the account given of the manufacture of nitro-glycerine and of the gelatine dynamites will be found more complete than in any similar work yet published in this country.

For many of the facts and figures contained in the chapter on Smokeless Powders I am indebted to (amongst others) the late Mr J.D. Dougall and Messrs A.C. Ponsonby and H.M. Chapman, F.C.S.; and for details with regard to Roburite to Messrs H.A. Krohn and W.J. Orsman, F.I.C. To these gentlemen my cordial thanks are due. Among the authorities which have been consulted in the general preparation of the work may be mentioned the *Journals* of the Chemical Society, the Society of Chemical Industry, the United States Naval Institute, and the Royal Artillery Institution. I have also referred to several volumes of the periodical publication *Arms and Explosives;* to various papers by Sir Frederick Abel, Bart., F.R.S., and General Wardell, R.A., on Gun-Cotton; to "Modern Artillery," by Capt. Lloyd, R.N., and A.G. Hadcock, R.A.; to the late Colonel Cundill's "Dictionary of Explosives"; as well as to the works of Messrs Eissler, Berthelot, and others.

The illustrations have been prepared chiefly from my own drawings. A few, however, have been taken (by permission) from the pages of *Arms and Explosives*, or from other sources which are acknowledged in the text.

P.G.S.

THE LABORATORY,

20 CULLUM STREET, E.C.

May 1896.

PREFACE TO THE SECOND EDITION.

In the preparation of the Second Edition of this work, I have chiefly made use of the current technical journals, especially of the *Journal of the Society of Chemical Industry*. The source of my information has in every case been acknowledged.

I am also indebted to several manufacturers of explosives for information respecting their special products — among others the New Explosives Company Ltd.; Messrs Curtis's and Harvey Ltd.; The Schultze Gunpowder Company Ltd.; and Mr W.D. Borland, F.I.C., of the E.C. Powder Company Ltd.

To my friend Mr A. Stanley Fox, F.C.S., of Faversham, my best thanks are also due for his help in many departments, and his kindness in pointing out several references.

The chapter on Smokeless Powders has been considerably enlarged and (as far as possible) brought up to date; but it has not always been possible to give the process of manufacture or even the composition, as these details have not, in several cases, been made public.

P. GERALD SANFORD.

LONDON, *June 1906.*

TABLE OF CONTENTS.

CHAPTER I.—INTRODUCTION.
The Nitro-Explosives—Substances that have been Nitrated—The Danger Area—Systems of Professors Lodge, Zenger, and Melsens for the Protection of Buildings from Lightning, &c.

CHAPTER II.—NITRO-GLYCERINE.
Properties of Nitro-Glycerine—Manufacture—Nitration—Separation—Washing and Filtering—Drying, Storing, &c.—The Waste Acids—Their Treatment— Nitric Acid Plants

CHAPTER III.—NITRO-CELLULOSE, &C.
Cellulose Properties—Discovery of Gun-Cotton—Properties of Gun-Cotton—Varieties of Soluble and Insoluble Gun-Cottons—Manufacture of Gun-Cotton—Dipping and SteepingWhirling Out the Acid—Washing, Boiling, Pulping, Compressing—The Waltham Abbey Process—Le Bouchet Process—Granulation of Gun-Cotton—Collodion-Cotton—Manufacture—Acid Mixture Used—Cotton Used, &c.—Nitrated Gun-Cotton—Tonite—Dangers in Manufacture of Gun-Cotton—Trench's Fire-Extinguishing Compound—Uses of Collodion-Cotton—Celluloid—Manufacture, &c.—Nitro-Starch, Nitro-Jute, and Nitro-Mannite

CHAPTER IV.—DYNAMITE.
Kieselguhr Dynamite—Classification of Dynamites—Properties and Efficiency of Ordinary Dynamite—Other forms of Dynamite—Gelatine and Gelatine Dynamites, Suitable Gun-Cotton for, and Treatment of—Other Materials Used—Composition of Gelignite—Blasting Gelatine—Gelatine Dynamite—Absorbing Materials—Wood Pulp—Potassium Nitrate, &c.—Manufacture, &c.—Apparatus Used—The Properties of the Gelatine Compounds

CHAPTER V.—NITRO-BENZOL, ROBURITE, BELLITE, PICRIC ACID, &c.

Explosives derived from Benzene—Toluene and Nitro-Benzene—Di- and Tri-nitro-Benzene—Roburite: Properties and Manufacture—Bellite: Properties, &c.—Securite—Tonite No. 3.—Nitro-Toluene—Nitro-Naphthalene—Ammonite—Sprengel's Explosives—Picric Acid—Picrates—Picric Powders—Melinite—Abel's Mixture—Brugère's Powders—The Fulminates—Composition, Formula, Preparation, Danger of, &c.—Detonators: Sizes, Composition, Manufacture—Fuses, &c.

THE FULMINATES.
Composition, Formula, Preparation, Danger of, &c.—Detonators: Sizes, Composition, Manufacture—Fuses, &c.

CHAPTER VI.—SMOKELESS POWDERS IN GENERAL.
Cordite—Axite—Ballistite—U.S. Naval Powder—Schultze's E.C. Powder—Indurite—Vielle Poudre—Walsrode and Cooppal Powders—Amberite—Troisdorf—B.N. Powder—Wetterin—Normal Powder—Maximite—Picric Acid Powders, &c. &c.

CHAPTER VII.—ANALYSIS OF EXPLOSIVES.
Kieselguhr Dynamite—Gelatine Compounds—Tonite—Cordite—Vaseline—Acetone—Scheme for Analysis of Explosives—Nitro-Cotton—Solubility Test—Non-Nitrated Cotton—Alkalinity—Ash and Inorganic Matter—Determination of Nitrogen—Lungé, Champion and Pellet's, Schultze-Tieman, and Kjeldahl's Methods—Celluloid—Picric Acid and Picrates—Resinous and Tarry Matters—Sulphuric Acid and Hydrochloric Acid and Oxalic Acid—Nitric Acid— Inorganic Impurities—General Impurities and Adulterations—Potassium Picrate, &c.—Picrates of the Alkaloids—Analysis of Glycerine—Residue—Silver Test—Nitration—Total Acid Equivalent—Neutrality—Free Fatty Acids—Combined Fatty Acids—Impurities—Oleic Acid—Sodium Chloride—Determination of Glycerine—Waste Acids—Sodium Nitrate—Mercury Fulminate—

Cap Composition—Table for Correction of Volumes of Gases, for Temperature and Pressure

CHAPTER VIII.—FIRING POINT OF EXPLOSIVES, HEAT TESTS, &C.

Horsley's Apparatus—Table of Firing Points—The Government Heat Test Apparatus, &c., for Dynamites, Nitro-Glycerine, Nitro-Cotton, and Smokeless Powders—Guttmann's Heat Test—Liquefaction and Exudation Tests—Page's Regulator for Heat Test Apparatus—Specific Gravities of Explosives—Will's Test for Nitro-Cellulose—Table of Temperature of Detonation, Sensitiveness, &c.

CHAPTER IX.—THE DETERMINATION OF THE RELATIVE STRENGTH OF EXPLOSIVES.

Effectiveness of an Explosive—High and Low Explosives—Theoretical Efficiency—M.M. Roux and Sarrau's Results—Abel and Noble's—Nobel's Ballistic Test—The Mortar—Pressure or Crusher Gauge—Calculation Volume of Gas Evolved, &c.—Lead Cylinders—The Foot-Pounds Machine—Noble's Pressure Gauge—Lieut. Walke's Results—Calculation of Pressure Developed by Dynamite and Gun-Cotton—McNab's and Ristori's Results of Heat Developed by the Explosion of Various Explosives—Composition of some of the Explosives in Common Use for Blasting, &c.

LIST OF ILLUSTRATIONS.

FRONTISPIECE—Danger Building showing Protecting Mounds. 1. Section of Nitro-Glycerine Conduit 2. Melsens System of Lightning Conductors 3. French System 4_a_ & 4_b_. English Government System 5. Upper Portion of Nitrator for Nitro-Glycerine 6. Small Nitrator 7. Nathan's Nitrator 8. Nitro-Glycerine Separator 9. Nitro-Glycerine Filtering Apparatus 10. Cotton-Waste Drier 11. Dipping Tank 12. Cooling Pits 13. Steeping Pot for Gun-Cotton 14. Hydro-Extractor or Centrifugal Drier 15_a_ & 15_b_. Gun-Cotton Beater 16_a_. Poacher for Pulping Gun-Cotton 16_b_. Plan of same 16_c_. Another form of Poacher 17 & 18. Compressed Gun-Cotton 19. Hydraulic Press 20. Thomson's Apparatus—Elevation 21. Elevation Plan 22. Trench's Safety Cartridge 23. Vessel used in Nitrating Paper 24. Cage ditto—White & Schupphaus' Apparatus 25. Do. do. do. 26 & 27. Nitrating Pot for Celluloid 28 & 29. Plunge Tank in Plan and Section 30. Messrs Werner, Pfleiderer & Perkins' Mixing Machine 31. M. 'Roberts' Mixing Machine for Blasting Gelatine 32. Plan of same 33. Cartridge Machine for Gelatines 34. Cartridge fitted with Fuse and Detonator 35. Gun-Cotton Primer 36. Electric Firing Apparatus 37. Metal Drum for Winding Cordite 38. Ten-Stranding 39. Curve showing relation between Pressures of Cordite and Black Powder, by Professor Vivian Lewes 40. Marshall's Apparatus for Moisture in Cordite 41. Lungé's Nitrometer 42. Modified do. 43. Horn's Nitrometer 44. Schultze-Tieman Apparatus for Determination of Nitrogen in Gun-Cotton 45. Decomposition Flask for Schultze-Tieman Method 46. Abel's Heat Test Apparatus 47. Apparatus for Separation of Nitro-Glycerine from Dynamite 48. Test Tube arranged for Heat Test 49. Page's Regulator 50. Do. showing Bye-Pass and Cut-off Arrangement 51. Will's Apparatus 52 & 53. Curves obtained 54. Dynamite Mortar 55. Quinan's Pressure Gauge 56. Steel Punch and Lead Cylinder for Use with Pressure Gauge 57. Micrometer Calipers for Measuring Thickness of Lead Cylinders 58. Section of Lead Cylinders before and after Explosion 59. Noble's Pressure Gauge 60. Crusher Gauge

NITRO-EXPLOSIVES.

CHAPTER I.

INTRODUCTORY.

The Nitro-Explosives—Substances that have been Nitrated—The Danger Area—
Systems of Professors Lodge, Zenger, and Melsens for the Protection of
Buildings from Lightning, &c.

The manufacture of the various nitro-explosives has made great advances during late years, and the various forms of nitro-compounds are gradually replacing the older forms of explosives, both for blasting purposes and also for propulsive agents, under the form of smokeless powders. The nitro-explosives belong to the so-called High Explosives, and may be defined as any chemical compound possessed of explosive properties, or capable of combining with metals to form an explosive compound, which is produced by the chemical action of nitric acid, either alone or mixed with sulphuric acid, upon any carbonaceous substance, whether such compound is mechanically mixed with other substances or not.[A]

[Footnote A: Definition given in Order of Council, No. 1, Explosives Act, 1875.]

The number of compounds and mixtures included under this definition is very large, and they are of very different chemical composition. Among the substances that have been nitrated are:—Cellulose, under various forms, e.g., cotton, lignin, &c.; glycerine, benzene, starch, jute, sugar, phenol, wood, straw, and even such substances as treacle and horse-dung. Some of these are not made upon the large scale, others are but little used. Those of most importance are nitro-glycerine and nitro-cellulose. The former enters into the composition of all dynamites, and several smokeless powders; and the second includes gun-cotton, collodion-cotton, nitrated wood, and the majority of the smokeless powders, which consist generally of nitro-cotton, nitro-lignin, nitro-jute, &c. &c., together with metallic nitrates, or nitro-glycerine.

The nitro-explosives consist generally of some organic substance in which the NO_2 group, known as nitryl, has been substituted in place of hydrogen.

Thus in glycerine,

|OH C_{3}H_{5}||OH, |OH

which is a tri-hydric alcohol, and which occurs very widely distributed as the alcoholic or basic constituent of fats, the hydrogen atoms are replaced by the NO_2 group, to form the highly explosive compound, nitro-glycerine. If one atom only is thus displaced, the mono-nitrate is formed thus,

|ONO_{2} C_{3}H_{5}||OH; |OH

and if the three atoms are displaced, $C_3H_5(ONO_2)_3$, or the tri- nitrate, is formed, which is commercial nitro-glycerine.

Another class, the nitro-celluloses, are formed from cellulose, $C_6H_{10}O_5$, which forms the groundwork of all vegetable tissues. Cellulose has some of the properties of the alcohols, and forms ethereal salts when treated with nitric and sulphuric acids. The hexa-nitrate, or gun-cotton, has the formula, $C_{12}H_{14}O_4(ONO_2)_6$; and collodion-cotton, pyroxylin, &c., form the lower nitrates, i.e., the tetra- and penta-nitrates. These last are soluble in various solvents, such as ether-alcohol and nitro-glycerine, in which the hexa-nitrate is insoluble. They all dissolve, however, in acetone and acetic ether.

The solution of the soluble varieties in ether-alcohol is known as collodion, which finds many applications in the arts. The hydrocarbon benzene, C_6H_6, prepared from the light oil obtained from coal-tar, when nitrated forms nitro-benzenes, such as mono-nitro-benzene, $C_6H_5NO_2$, and di-nitro-benzene, $C_6H_4(NO_2)_2$, *in which one and two atoms are replaced by the*

NO_2 group. The latter of these compounds is used as an explosive, and enters into the composition of such well-known explosives as roburite, &c. The presence of nitro groups in a substance increases the difficulty of further nitration, and in any case not more than three nitro groups can be introduced into an aromatic compound, or the phenols. All aromatic compounds with the general formula, $C_6H_4X_2$, give, however, three series. They are called ortho, meta, or para compounds, depending upon the position of NO_2 groups introduced.

Certain regularities have been observed in the formation of nitro-compounds. If, for example, a substance contains alkyl or hydroxyl groups, large quantities of the para compound are obtained, and very little of the ortho. The substitution takes place, however, almost entirely in the meta position, if a nitro, carboxyl, or aldehyde group be present. Ordinary phenol, $C_6H_5.OH$, gives para- and ortho-nitro-phenol; toluene gives para- and ortho-nitro-toluene; but nitro-benzene forms meta-di-nitro- benzene and benzoic acid, meta-nitro-benzoic acid.[A]

[Footnote A: "Organic Chemistry," Prof. Hjelt. Translated by J.B. Tingle,
Ph.D.]

If the graphic formula of benzene be represented thus (No. 1), then the positions 1 and 2 represent the ortho, 1 and 3 the meta, and 1 and 4 the para compounds. When the body phenol, $C_6H_5.OH$, is nitrated, a compound is formed known as tri-nitro-phenol, or picric acid, $C_6H_2(NO_2)_3OH$, *which is used very extensively as an explosive, both as picric acid and in the form of picrates. Another nitro body that is used as an explosive is nitro-naphthalene,* $C_{10}H_6(NO_2)_2$, *in roburite, securite, and other explosives of this class. The hexa-nitro- mannite,* $C_6H_8(ONO_2)_6$, is formed

[Illustration: No. 1]

[Illustration: META-DINITRO-BENZENE No.2]

by treating a substance known as mannite, $C_6H_8(OH)_6$, an alcohol formed by the lactic acid fermentation of sugar and closely related to the sugars, with nitric and sulphuric acids. It is a solid substance, and very explosive; it contains 18.58 per cent. of nitrogen.

Nitro-starch has also been used for the manufacture of an explosive. Muhlhauer has described (*Ding. Poly. Jour.*, 73, 137-143) three nitric ethers of starch, the tetra-nitro-starch, $C_{12}H_{16}O_{6}(ONO_{2})_{4}$, the penta- and hexa-nitro-starch. They are formed by acting upon potato starch dried at 100° C. with a mixture of nitric and sulphuric acids at a temperature of 20° to 25° C. Rice starch has also been used in its production. Muhlhauer proposes to use this body as a smokeless powder, and to nitrate it with the spent mixed acids from the manufacture of nitro-glycerine. This substance contains from 10.96 to 11.09 per cent. of nitrogen. It is a white substance, very stable and soluble even in cold nitro-glycerine.

The explosive bodies formed by the nitration of jute have been studied by Messrs Cross and Bevan. and also by Mühlhäuer. The former chemists give jute the formula $C_{12}H_{18}O_{9}$, and believe that its conversion into a nitro-compound takes place according to the equation—

$$C_{12}H_{18}O_{9} + 3HNO_{3} = 3H_{2}O + C_{12}H_{15}O_{6}(NO_{3})_{3}.$$

This is equivalent to a gain in weight of 44 per cent. for the tri-nitrate, and 58 per cent. for the tetra-nitrate. The formation of the tetra-nitrate appears to be the limit of nitration of jute fibre. Messrs Cross and Bevan say, "In other words, if we represent the ligno-cellulose molecule by a C_{12} formula, it will contain four hydroxyl (OH) groups, or two less than cellulose similarly represented." It contains 11.5 per cent. of nitrogen. The jute nitrates resemble those of cellulose, and are in all essential points nitrates of ligno-cellulose.

Nitro-jute is used in the composition of the well-known Cooppal Smokeless Powders. Cross and Bevan are of opinion that there is no very obvious advantage in the use of lignified textile fibres as raw materials for explosive nitrates, seeing that a number of raw materials containing cellulose (chiefly as cotton) can be obtained at from £10 to £25 a ton, and yield also 150 to 170 per cent. of explosive material when nitrated (whereas jute only gives 154.4 per cent.), and are in many ways superior to the products obtained from jute. Nitro-lignin, or nitrated wood, is, however, largely used in the composition of a good many of the smokeless powders, such as Schultze's, the Smokeless Powder Co.'s products, and others.

~The Danger Area.~ — That portion of the works that is devoted to the actual manufacture or mixing of explosive material is generally designated by the term "danger area," and the buildings erected upon it are spoken of as "danger buildings." The best material of which to construct these buildings is of wood, as in the event of an explosion they will offer less resistance, and will cause much less danger than brick or stone buildings. When an explosion of nitro-glycerine or dynamite occurs in one of these buildings, the sides are generally blown out, and the roof is raised some considerable height, and finally descends upon the blown-out sides. If, on the other hand, the same explosion had occurred in a strong brick or stone building, the walls of which would offer a much larger resistance, large pieces of brickwork would probably have been thrown for a considerable distance, and have caused serious damage to surrounding buildings.

It is also a very good plan to surround all danger buildings with mounds of sand or earth, which should be covered with turf, and of such a height as to be above the roof of the buildings that they are intended to protect (see frontispiece).[A] These mounds are of great value in confining the force of the explosion, and the sides of the buildings being thrown against them are prevented from travelling any distance. In gunpowder works it is not unusual to surround the danger buildings with trees or dense underwood instead of mounds. This would be of no use in checking the force of explosion of the high explosives, but has been found a very useful precaution in the case of gunpowder.

[Footnote A: At the Baelen Factory, Belgium, the danger buildings are erected on a novel plan. They are circular in ground plan and lighted entirely from the roof by means of a patent glass having wire-netting in it, and which it is claimed will not let a splinter fall, even if badly cracked. The mounds are then erected right up against the walls of the building, exceeding them in height by several metres. For this method of construction it is claimed that the force exerted by an explosion will expand itself in a vertical direction ("Report on Visits to Certain Explosive Factories," H.M. Inspectors, 1905).]

In Great Britain it is necessary that all danger buildings should be a specified distance apart; a license also must be obtained. The application for a license must give a plan (drawn to scale) of the proposed factory or magazine, and the site, its boundaries, and surroundings, and distance the building will be from any other buildings or works, &c., also the character, and construction of all the mounds, and nature of the processes to be carried on in the factory or building.[A]

[Footnote A: Explosives Act, 38 Vict. ch. 17.]

[Illustration: FIG. 1.—SECTION OF NITRO-GLYCERINE CONDUIT. *a*, lid; *b*, lead lining; *c*, cinders.]

The selection of a site for the danger area requires some attention. The purpose for which it is required, that is, the kind of explosive that it is intended to manufacture, must be taken into consideration. A perfectly level piece of ground might probably be quite suitable for the purpose of erecting a factory for the manufacture of gun-cotton or gunpowder, and such materials, but would be more or less unsuitable for the manufacture of nitro-glycerine, where a number of buildings are required to be upon different levels, in order to allow of the flow of the liquid nitro- glycerine from one building to another through a system of conduits. These conduits (Fig. 1), which are generally made of wood and lined with lead, the space between the woodwork and the lead lining, which is generally some 4 or 5 inches, being filled with cinders, connect the various buildings, and should slope gently from one to the other. It is also desirable that, as far as possible, they should be protected by earthwork banks, in the same way as the danger buildings themselves. They should also be provided with covers, which should be whitewashed in hot weather.

A great deal of attention should be given to these conduits, and they should be very frequently inspected. Whenever it is found that a portion of the lead lining requires repairing, before cutting away the lead it should be very carefully washed, for several feet on either side of the portion that it is intended to remove, with a solution of caustic soda or potash dissolved in methylated spirit and water, and afterwards with water alone. This decomposes the nitroglycerine forming glycerine and potassium nitrate. It will be found that the mixed acids attack the lead rather quickly, forming sulphate and nitrate of lead, but chiefly the former. It is on this account that it has been proposed to use pipes made of guttapercha, but the great drawback to their use is that in the case of anything occurring inside the pipes, such as the freezing of the nitro- glycerine in winter, it is more difficult to find it out, and the condition of the inside cannot be seen, whereas in the case of wooden conduits it is an easy matter to lift the lids along the whole length of the conduit.

The buildings which require to be connected by conduits are of course those concerned with the manufacture of nitro-glycerine. These buildings are—(1) The nitrating house; (2) the separating house; (3) the filter house; (4) the secondary separator; (5) the deposit of washings; (6) the settling or precipitation house; and each of these buildings must be on a level lower than the preceding one, in order that the nitro-glycerine or acids may flow easily from one building to the next. These buildings are, as far as possible, best placed together, and away from the other danger buildings, such as the cartridge huts and dynamite mixing houses, but this is not essential.

All danger buildings should be protected by a lightning conductor, or covered with barbed wire, as suggested by Professor Sir Oliver J. Lodge, F.R.S., Professors Zenger, of Prague, and Melsens, of Brussels, and everything possible should be done to keep them as cool as possible in the summer. With this object they should be made double, and the intervening space filled with cinders. The roof also should be kept whitewashed, and the windows painted over thinly with white paint. A thermometer should be suspended in every house. It is very essential that the floors of all these buildings should be washed every day before the work-people leave. In case any nitro-glycerine is spilt upon the floors, after sponging it up as far as possible, the floor should be washed with an alcoholic solution of soda or potash to decompose the nitro-glycerine, which it does according to the equation[A]—

$$C_3H_5(NO_3)_3 + 3KOH = C_3H_8O_3 + 3KNO_3.$$

[Footnote A: See also Berthelot, *Comptes Rendus*, 1900, 131[12], 519- 521.]

Every one employed in the buildings should wear list or sewn leather shoes, which of course must be worn in the buildings only. The various houses should be connected by paths laid with cinders, or boarded with planks, and any loose sand about the site of the works should be covered over with turf or cinders, to prevent its blowing about and getting into the buildings. It is also of importance that stand pipes should be placed about the works with a good pressure of water, the necessary hose being kept in certain known places where they can be at once got at in the case of fire, such as the danger area laboratory, the foreman's office, &c. It is also desirable that the above precautions against fire should be tested once a week. With regard to the heating of the various buildings in the winter, steam pipes only should be used, and should be brought from a boiler-house outside the danger area, and should be covered with kieselguhr or fossil meal and tarred canvas. These pipes may be supported upon poles. A stove of some kind should be placed in the corner of each building, but it must be entirely covered in with woodwork, and as small a length of steam pipes should be within the building as possible.

In the case of a factory where nitro-glycerine and dynamite are manufactured, it is necessary that the work-people should wear different clothes upon the danger area than usual, as they are apt to become impregnated with nitro-glycerine, and thus not very desirable or safe to wear outside the works. It is also necessary that these clothes should not contain any pockets, as this lessens the chance of matches or steel implements being taken upon the danger area. Changing houses, one for the men, and another for the girls, should also be provided. The tools used upon the danger area should, whenever the building is in use, or contains explosives, be made of phosphor bronze or brass, and brass nails or wooden pegs should be used in the construction of all the buildings.

[Illustration: FIG. 2. — MELSENS SYSTEM OF LIGHTNING CONDUCTORS.]

~Lightning Conductors.~ — The Explosive Substances Act, 38 Vict. ch. 17, clause 10, says, "Every factory magazine and expense magazine in a factory, and every danger building in a magazine, shall have attached thereto a sufficient lightning conductor, unless by reason of the construction by excavation or the position of such magazine or building, or otherwise, the Secretary of State considers a conductor unnecessary, and every danger building in a factory shall, if so required by the Secretary of State, have attached thereto a sufficient lightning conductor."

The exact form of lightning conductor most suitable for explosive works and buildings has not yet been definitely settled. Lightning-rod engineers favour what is known as the Melsens system, due to Professor Melsens, of Brussels, and Professor Zenger, of Prague, but first suggested by the late Professor Clerk-Maxwell. In a paper read before the British Association, Clerk-Maxwell proposed to protect powder-magazines from the effects of lightning by completely surrounding or encasing them with sheet metal, or a cage of metallic conductors. There were, however, several objections to his system as he left it.

Professor Melsens[A] has, while using the idea, made several important alterations. He has multiplied the terminals, the conductors, and the earth-connections. His terminals are very numerous, and assume the form of an aigrette or brush with five or seven points, the central point being a little higher than the rest, which form with it an angle of 45°. He employs for the most part galvanised-iron wire. He places all metallic bodies, if they are of any considerable size, in communication with the conducting system in such a manner as to form closed metallic circuits. His system is illustrated in Fig. 2, taken from *Arms and Explosives*.

[Footnote A: Belgian Academy of Science.]

This system is a near approximation to J.C. Maxwell's cage. The system was really designed for the protection of powder-magazines or store buildings placed in very exposed situations. Zenger's system is identical with that of Melsens, and has been extensively tried by the Austrian military authorities, and Colonel Hess has reported upon the absolute safety of the system.

[Illustration: Fig. 3.—FRENCH SYSTEM OF LIGHTNING CONDUCTORS.]

The French system of protecting powder-magazines is shown in Fig. 3, where there are no brush terminals or aigrettes. The French military authorities also protect magazines by erecting two or more lightning-rods on poles of sufficient height placed close to, but not touching, the walls of the magazine. These conductors are joined below the foundations and earthed as usual.

In the instructions issued by the Government, it is stated that the lightning-rods placed upon powder-mills should be of such a height, and so situated, that no danger is incurred in igniting the powder-dust in the air by the lightning discharge at the pointed rod. In such a case a fork or aigrette of five or more points should invariably be used in place of a single point.

[Illustration: FIG. 4_a_.—GOVERNMENT SYSTEM OF LIGHTNING CONDUCTORS FOR LARGE BUILDINGS.]

[Illustration: FIG. 4_b_.—GOVERNMENT SYSTEM OF LIGHTNING CONDUCTORS FOR SMALL BUILDINGS.]

In Fig. 4 (*a* and *b*) is shown the Government method for protecting buildings in which explosives are made or stored. Multiple points or aigrettes would be better. Lord Kelvin and Professor Melsens favour points, and it is generally admitted that lightning does not strike buildings at a single point, but rather in a sheet; hence, in such cases, or in the event of the globular form being assumed by the lightning, the aigrette will constitute a much more effective protection than a single point. As to the spacing of conductors, they may, even on the most important buildings, be spaced at intervals of 50 feet. There will then be no point on the building more than 25 feet from the conductor. This "25-feet rule" can be adhered to with advantage in all overground buildings for explosives.

Underground magazines should, whenever possible, also be protected, because, although less exposed than overground buildings, they frequently contain explosives packed in metal cases, and hence would present a line of smaller electrical resistance than the surrounding earth would offer to the lightning. The conductor should be arranged on the same system as for overground buildings, but be applied to the surface of the ground over the magazines.

In all situations where several conductors are joined in one system, the vertical conductors should be connected both at the top and near the ground line. The angles and the prominent portions of a building being the most liable to be struck, the conductors should be carried over and along these projections, and therefore along the ridges of the roof. The conductors should be connected to any outside metal on the roofs and walls, and specially to the foot of rain-water pipes.

All the lightning conductors should be periodically tested, to see that they are in working condition, at least every three months, according to Mr Richard Anderson. The object of the test is to determine the resistance of the earth-connection, and to localise any defective joints or parts in the conductors. The best system of testing the conductors is to balance the resistance of each of the earths against the remainder of the system, from which the state of the earths may be inferred with sufficient accuracy for all practical purposes.

Captain Bucknill, R.E., has designed an instrument to test resistance which is based on the Post Office pattern resistance coil, and is capable of testing to approximate accuracy up to 200 ohms, and to measure roughly up to 2,000 ohms. Mr R. Anderson's apparatus is also very handy, consisting of a case containing three Leclanché cells, and a galvanometer with a "tangent" scale and certain standard resistances. Some useful articles on the protection of buildings from lightning will be found in *Arms and Explosives*, July, August, and September 1892, and by Mr Anderson, Brit. Assoc., 1878-80.

~Nitro-Glycerine.~—One of the most powerful of modern explosive agents is nitro-glycerine. It is the explosive contained in dynamite, and forms the greater part of the various forms of blasting gelatines, such as gelatine dynamite and gelignite, both of which substances consist of a mixture of gun-cotton dissolved in nitro-glycerine, with the addition of varying proportions of wood-pulp and saltpetre, the latter substances acting as absorbing materials for the viscid gelatine. Nitro-glycerine is also largely used in the manufacture of smokeless powders, such as cordite, ballistite, and several others.

Nitro-glycerol, or glycerol tri-nitrate, was discovered by Sobrero in the year 1847. In a letter written to M. Pelouse, he says, "when glycerol is poured into a mixture of sulphuric acid of a specific gravity of 1.84, and of nitric acid of a gravity of 1.5, which has been cooled by a freezing mixture, that an oily liquid is formed." This liquid is nitro-glycerol, or nitro-glycerine, which for some years found no important use in the arts, until the year 1863, when Alfred Nobel first started a factory in Stockholm for its manufacture upon a large scale; but on account of some serious accidents taking place, its use did not become general.

It was not until Nobel conceived the idea (in 1866) of absorbing the liquid in some absorbent earth, and thus forming the material that is now known as dynamite, that the use of nitro-glycerine as an explosive became general.

Among those who improved the manufacture of nitro-glycerine was Mowbray, who, by using pure glycerine and nitric acid free from nitrous acid, made very great advances in the manufacture. Mowbray was probably the first to use compressed air for the purpose of keeping the liquids well agitated during the process of nitration, which he conducted in earthenware pots, each containing a charge of 17 lbs. of the mixed acids and 2 lbs. of glycerol.

A few years later (1872), MM. Boutnny and Faucher, of Vonges,[A] proposed to prepare nitro-glycerine by mixing the sulphuric acid with the glycerine, thus forming a sulpho-glyceric acid, which was afterwards mixed with a mixture of nitric and sulphuric acids. They claimed for this method of procedure that the final temperature is much lower. The two mixtures are mixed in the proportions—Glycerine, 100; nitric acid, 280; and sulphuric acid, 600. They state that the rise of temperature upon mixing is limited from 10° to 15° C.; but this method requires a period of twenty-four hours to complete the nitration, which, considering the danger of keeping the nitro-glycerine in contact with the mixed acids for so long, probably more than compensates for the somewhat doubtful advantage of being able to perform the nitration at such a low temperature. The Boutnny process was in operation for some time at Pembrey Burrows in Wales, but after a serious explosion the process was abandoned.

[Footnote A: *Comptes Rendus*, 75; and Desortiaux, "Traité sur la Poudre," 684-686.]

Nitro-glycerine is now generally made by adding the glycerine to a mixture of sulphuric and nitric acids. The sulphuric acid, however, takes no part in the reaction, but is absolutely necessary to combine with the water that is formed by the decomposition, and thus to keep up the strength of the nitric acid, otherwise lower nitrates of glycerine would be formed that are soluble in water, and which would be lost in the subsequent process of washing to which the nitro-compound is subjected, in order to remove the excess of acids, the retention of which in the nitro-glycerol is very dangerous. Nitroglycerol, which was formerly considered to be a nitro-substitution compound of glycerol, was thought to be formed thus —

$$C_3H_8O_3 + 3HNO_3 = C_3H_5(NO_2)_3O_3 + 3H_2O;$$

but more recent researches rather point to its being regarded as a nitric ether of glycerol, or glycerine, and to its being formed thus —

$$C_3H_8O_3 + 3\ HNO_3 = C_3H_5(NO_3)_3 + 3H_2O.$$
$$92227$$

The formula of glycerine is $C_3H_8O_8$, or $C_3H_5 \begin{array}{|l} OH \\ OH \\ OH \end{array}$

and that of the mono-nitrate of glycerine, $C_3H_5 \begin{array}{|l} ONO_2 \\ OH \\ OH \end{array}$

and of the tri-nitrate or (nitro-glycerine), $C_3H_5 \begin{array}{|l} ONO_2 \\ ONO_2 \\ ONO_2 \end{array}$

that is, the three hydrogens of the semi-molecules of hydroxyl in the glycerine have been replaced by the NO_2 group.

In the manufacture upon the large scale, a mixture of three parts by weight of nitric acid and five parts of sulphuric acid are used. From the above equation it will be seen that every 1 lb. of glycerol should give 2.47 lbs. of nitro-glycerol $((227+1)/92 = 2.47)$, but in practice the yield is only about 2 lbs. to 2.22, the loss being accounted for by the unavoidable formation of some of the lower nitrate, which dissolves in water, and is thus washed away, and partly perhaps to the presence of a little water (or other non-nitrable matter) in the glycerine, but chiefly to the former, which is due to the acids having become too weak.

CHAPTER II.

MANUFACTURE OF NITRO-GLYCERINE.

Properties of Nitro-Glycerine — Manufacture of Nitro-Glycerine — Nitration —
The Nathan Nitrator — Separation — Filtering and Washing — The Waste Acids —
Treatment of the Waste Acid from the Manufacture of Nitro-Glycerine and
Gun-Cotton.

~Properties of Nitro-Glycerine.~ — Nitro-glycerol is a heavy oily liquid of specific gravity 1.6 at 15° C., and when quite pure is colourless. The commercial product is a pale straw yellow, but varies much according to the purity of the materials used in its manufacture. It is insoluble in water, crystallises at 10.5° C., but different commercial samples behave very differently in this respect, and minute impurities prevent or delay crystallisation. Solid nitro-glycerol[A] melts at about 12° C., but requires to be exposed to this temperature for some time before melting. The specific gravity of the solid form is 1.735 at +10° C.; it contracts one-twelfth of its volume in solidifying. Beckerheim[B] gives the specific heat as 0.4248 between the temperatures of 9.5° and 9.8° C., and L. de Bruyn gives the boiling point as above 200°.

[Footnote A: Di-nitro-mono chlorhydrin, when added to nitro-glycerine up to 20 per cent., is said to prevent its freezing.]

[Footnote B: *Isb., Chem. Tech.*, 22, 481-487. 1876.]

Nitro-glycerine has a sweet taste, and causes great depression and vertigo. It is soluble in ether, chloroform, benzene, glacial acetic acid, and nitro-benzene, in 1.75 part of methylated spirit, very nearly insoluble in water, and practically insoluble in carbon bisulphide. Its formula is $C_3H_5(NO_3)_3$, and molecular weight 227.

When pure, it may be kept any length of time without decomposition. Berthelot kept a sample for ten years, and Mr G. M'Roberts, of the Ardeer Factory, for nine years, without their showing signs of decomposition; but if it should contain the smallest trace of free acid, decomposition is certain to be started before long. This will generally show itself by the formation of little green spots in the gelatine compounds, or a green ring upon the surface of liquid nitro-glycerine. Sunlight will often cause it to explode; in fact, a bucket containing some water that had been used to wash nitroglycerine, and had been left standing in the sun, has in our experience been known to explode with considerable force. Nitroglycerine when pure is quite stable at ordinary temperatures, and samples have been kept for years without any trace of decomposition. It is very susceptible to heat, and even when quite pure will not stand a temperature of 100° C. for a longer period than a few hours, without undergoing decomposition. Up to a temperature of 45° C., however, properly made and purified nitro- glycerine will remain unchanged almost indefinitely. The percentage composition of nitroglycerine is as follows:—

Found. Theory for $C_3H_5(NO_2)_3$.

Carbon 15.62 15.86 per cent.
Hydrogen 2.40 2.20 "
Nitrogen 17.90 18.50 "
Oxygen ... 63.44 "

The above analysis is by Beckerheim. Sauer and Adou give the nitrogen as 18.35 to 10.54 per cent. by Dumas' method; but I have never found any difficulty in obtaining percentages as high as 18.46 by the use of Lunge's nitrometer. The decomposition products by explosion are shown by the following equation—

$2C_3H_5(NO_3)_3 = 6CO_2 + 5H_2O + 6N + O$;

that is, it contains an excess of 3.52 per cent. of oxygen above that required for complete combustion; 100 grms. would be converted into—

Carbonic Acid (CO_2) 58.15 per cent.
Water 19.83 "

Oxygen 3.52 per cent.
Nitrogen 18.50 "

The volume of gases produced at 0° and 760 mm., calculated from the above, is 714 litres per kilo, the water being taken as gaseous. Nitro-glycerine is decomposed differently if it is ignited as dynamite (i.e., kieselguhr dynamite), and if the gases are allowed to escape freely under a pressure nearly equal to that of the atmosphere. Sarrau and Vieille obtained under these conditions, for 100 volumes of gas—

NO 48.2 per cent.
CO 35.9 "
CO_2 12.7 "
H 1.6 per cent.
N 1.3 "
CH_4 0.3 "

These conditions are similar to those under which a mining charge, simply ignited by the cap, burns away slowly under a low pressure (i.e., a miss fire). In a recent communication, P.F. Chalon (*Engineering and Mining Journal*, 1892) says, that in practice nitro-glycerine vapour, carbon monoxide, and nitrous oxide, are also produced as the result of detonation, but he attributes their formation to the use of a too feeble detonator.

Nitro-glycerine explodes very violently by concussion. It may be burned in an open vessel, but if heated above 250° C. it explodes. Professor C.E. Munroe gives the firing point as 2O3°-2O5° C., and L. de Bruyn[A] states its boiling point as 185°. He used the apparatus devised by Horsley. The heat of formation of nitro-glycerine, as deduced from the heat of combustion by M. Longuinine, is 432 calories for 1 grm.; and the heat of combustion equals 1,576 cals. for 1 grm. In the case of nitro-glycerine the heat of total combustion and the heat of complete decomposition are interchangeable terms, since it contains an excess of oxygen. According to Dr W.H. Perkin, F.R.S.,[B] the magnetic rotation of nitro-gylcerine is 5,407, and that of tri-methylene nitrate, 4.769 (diff. = .638). Dr Perkin says: "Had nitro-glycerine contained its nitrogen in any other combination with oxygen than as -O-NO_{2}, as it might if its constitution had been represented as C_{3}H_{2}(NO_{2}){3}(OH){3}, the rotation when compared with propyl nitrate (4.085) would be abnormal."

[Footnote A: *Jour. Soc. Chem. Ind.*, June 1896, p. 471.]

[Footnote B: *Jour. Chem. Soc.*, W.H. Perkin, 1889, p. 726.]

The solubility of nitro-glycerine in various solvents has been investigated by A.H. Elliot; his results may be summarised as follows:—

Solvent.	Cold.	Warm.
Water	Insoluble	Slightly soluble
Alcohol, absolute	Soluble	Soluble
" 93%	"	"
" 80%	Slowly soluble	"
" 50%	Insoluble	Slightly soluble
Methyl alcohol	Soluble	Soluble
Amyl "	"	"
Ether, ethylic	"	"
" acetic	"	"
Chloroform	"	"
Acetone	"	"
Sulphuric acid (1.845)	"	"
Nitric acid (1.400)	Slowly soluble	"
Hydrochloric acid (1.200)	Insoluble, decomposed	Slowly soluble
Acetic acid, glacial	Soluble	Soluble
Carbolic acid	"	"
Astral oil	Insoluble	Insoluble
Olive "	Soluble	Soluble
Stearine oil	"	"
Mineral jelly	Insoluble	Insoluble
Glycerine	"	"
Benzene	Soluble	Soluble
Nitro-benzene	"	"
Toluene	"	"
Carbon bi-sulphide	Insoluble	Slightly affected
Turpentine	"	Soluble
Petroleum naphtha, 71°-76° B.	"	Insoluble
Caustic soda (1:10 solution)	Insoluble.	Insoluble.
Borax, 5% solution	"	"
Ammonia (.980)	"	" slightly affected.
Ammonium sulph-hydrate	Insoluble, sulphur separates	Decomposed.
Iron sulphate solution	Slightly affected	Affected.
Iron chloride (1.4 grm. Fe to 10 c.c. $N_{2}O$)	Slowly affected	Decomposed.
Tin chloride	Slightly affected	Affected.

Many attempts have been made to prepare nitro-glycerine explosives capable of withstanding comparatively low temperatures without freezing, but no satisfactory solution of the problem has been found. Among the substances that have been proposed and used with more or less success, are nitro- benzene, nitro-toluene, di-nitro-mono-chlorhydrine, solid nitro derivatives of toluene,[A] are stated to lower the freezing point of nitro-glycerine to -20°C. without altering its sensitiveness and stability. The subject has been investigated by S. Nauckhoff,[B] who states that nitroglycerine can be cooled to temperatures (-40° to -50° C.) much below its true freezing point, without solidifying, by the addition of various substances. When cooled by means of a mixture of solid carbon, dioxide, and ether, it sets to a glassy mass, without any perceptible crystallisation. The mass when warmed to 0°C. first rapidly liquefies and then begins to crystallise. The true freezing point of pure nitro-glycerine was found to be 12.3°C. The technical product, owing to the presence of di-nitro-glycerine, freezes at 10.5° C. According to Raoult's law, the lowering of the freezing point caused by m grms. of a substance with the molecular weight M, when dissolved in 100 grms. of the solvent, is expressed by the formula: [Delta] = E(m/M), where E is a constant characteristic for the solvent in question. The value of E for nitro- glycerine was found to be 70.5 when calculated, according to Van't Hoff's formula, from the melting point and the latent heat of fusion of the substance. Determinations of the lowering of the freezing point of nitro- glycerine by additions of benzene, nitro-benzene, di-nitro-benzene, tri- nitro-benzene, p.-nitro-toluene, o.-nitro-toluene, di-nitro-toluene, naphthalene, nitro-naphthalene, di-nitro-naphthalene, ethyl acetate, ethyl nitrate, and methyl alcohol, gave results agreeing fairly well with Raoult's formula, except in the case of methyl alcohol, for which the calculated lowering of the freezing point was greater than that observed, probably owing to the formation of complex molecules in the solution. The results show that, in general, the capacity of a substance to lower the freezing point of nitro-glycerine depends, not upon its freezing point, or its chemical composition or constitution, but upon its molecular weight. Nauckhoff states that a suitable substance for dissolving in nitro- glycerine, in order to lower the freezing point of the latter, must have a relatively low molecular weight, must not appreciably diminish the explosive power and stability of the explosive, and must not be easily volatile at relatively high atmospheric temperatures; it should, if possible, be a solvent of nitro-cellulose, and in every case must not have a prejudicial influence on the gelatinisation of the nitro-cellulose.

[Footnote A: Eng. Pat. 25,797, November 1904.]

[Footnote B: *Z. Angew. Chem.*, 1905, 18, 11-22, 53-60.]

~Manufacture of Nitro-Glycerine.~ — Nitro-glycerine is prepared upon the manufacturing scale by gradually adding glycerine to a mixture of nitric and sulphuric acids of great strength. The mixed acids are contained in a lead vessel, which is kept cool by a stream of water continually passing through worms in the interior of the nitrating vessel, and the glycerine is gradually added in the form of a fine stream from above. The manufacture can be divided into three distinct operations, viz., nitration, separation, and washing, and it will be well to describe these operations in the above order.

~Nitration.~ — The most essential condition of nitrating is the correct composition and strength of the mixed acids. The best proportions have been found to be three parts by weight of nitric acid of a specific gravity 1.525 to 1.530, and containing as small a portion of the oxides of nitrogen as possible, to five parts by weight of sulphuric acid of a specific gravity of 1.840 at 15° C., and about 97 per cent. of mono- hydrate. It is of the very greatest importance that the nitric acid should be as strong as possible. Nothing under a gravity of 1.52 should ever be used even to mix with stronger acid, and the nitration will be proportional to the strength of the acid used, provided the sulphuric acid is also strong enough. It is also of great importance that the oxides of nitrogen should be low, and that they should be kept down to as low as 1 per cent., or even lower. It is also very desirable that the nitric acid should contain as little chlorine as possible. The following is the analysis of a sample of nitric acid, which gave very good results upon the commercial scale: — Specific gravity, 1.525, N_2O_4, 1.03 per cent.; nitric acid (HNO_3), 95.58 per cent.

The amount of real nitric acid (mono-hydrate) and the amount of nitric peroxide present in any sample should always be determined before it is used for nitrating purposes. The specific gravity is not a sufficient guide to the strength of the acid, as an acid having a high gravity, due to some 3 or 4 per cent of nitric oxides in solution, will give very poor nitration results. A tenth normal solution of sodium hydroxide (NaOH), with phenol-phthalein as indicator, will be found the most convenient method of determining the total acid present. The following method will be found to be very rapid and reliable:—Weigh a 100 c.c. flask, containing a few cubic centimetres of distilled water, and then add from a pipette 1 c.c. of the nitric acid to be examined, and reweigh (this gives the weight of acid taken). Now make up to 100 c.c. at 15° C.; shake well, and take out 10 c.c. with a pipette; drain into a small Erlenmeyer flask, and add a little of the phenol-phthalein solution, and titrate with the tenth normal soda solution.

The nitric peroxide can be determined with a solution of potassium permanganate of N/10 strength, thus: Take a small conical flask, containing about 10 c.c. of water, and add from a burette 10 to 16 c.c. of the permanganate solution; then add 2 c.c. of the acid to be tested, and shake gently, and continue to add permanganate solution as long as it is decolourised, and until a faint pink colour is permanent.

Example. N/10 permanganate 3.16 grms. per litre, 1 c.c. = O.0046 grm. N_2O_4, 2 c.c. of sample of acid specific gravity 1.52 = 3.04 grms. taken for analysis. Took 20 c.c. permanganate solution, O.0046 x 20 =.092 grm. N_2O_4, and (.092 x 100)/3.04 = 3.02 per cent. N_2O_4. The specific gravity should be taken with an hydrometer that gives the specific gravity directly, or, if preferred, the 2 c.c. of acid may be weighed.

A very good method of rapidly determining the strength of the sulphuric acid is as follows:—Weigh out in a small weighing bottle, as nearly as possible, 2.45 grms. This is best done by running in 1.33 c.c. of the acid (1.33 x 1.84 = 2.447). Wash into a large Erlenmeyer flask, carefully washing out the bottle, and also the stopper, &c. Add a drop of phenol- phthalein solution and titrate, with a half normal solution of sodium hydrate (use a 100 c.c. burette). Then if 2.45 grms. exactly have been taken, the readings on the burette will equal percentages of H_2SO_4 (mono-hydrate) if not, calculate thus:—2.444 grms. weighed, required 95.4 c.c. NaOH. Then—

2.444 : 95.4 :: 2.45 : x = 95.64 per cent. H_2SO_4.

It has been proposed to free nitric acid from the oxides of nitrogen by blowing compressed air through it, and thus driving the gases in solution out. The acid was contained in a closed lead tank, from which the escaping fumes were conducted into the chimney shaft, and on the bottom of which was a lead pipe, bent in the form of a circle, and pierced with holes, through which the compressed air was made to pass; but the process was not found to be of a very satisfactory nature, and it is certainly better not to allow the formation of these compounds in the manufacture of the acid in the first instance. Another plan, however, is to heat the acid gently, and thus drive out the nitrous gases. Both processes involve loss of nitric acid.

Having obtained nitric and sulphuric acids as pure as possible, the next operation is to mix them. This is best done by weighing the carboys in which the acids are generally stored before the acids are drawn off into them from the condensers, and keeping their weights constantly attached to them by means of a label. It is then a simple matter to weigh off as many carboys of acid as may be required for any number of mixings, and subtract the weights of the carboys. The two acids should, after being weighed, be poured into a tank and mixed, and subsequently allowed to flow into an acid egg or montjus, to be afterwards forced up to the nitrating house in the danger area. The montjus or acid egg is a strong cast-iron tank, of either an egg shape, or a cylinder with a round end. If of the former shape, it would lie on its side, and upon the surface of the ground, and would have a manhole at one end, upon which a lid would be strongly bolted down; but if of the latter shape, the lid, of course, is upon the top, and the montjus itself is let into the ground. In either case, the principle is the same. One pipe, made of stout lead, goes to the bottom, and another just inside to convey the compressed air, the acids flowing away as the pressure is put on, just as blowing down one tube of an ordinary wash- bottle forces the water up the other tube to the jet. The pressure necessarily will, of course, vary immensely, and will depend upon the height to which the acid has to be raised and the distance to be traversed.

The mixed acids having been forced up to the danger area, and to a level higher than the position of the nitrating house, should, before being used, be allowed to cool, and leaden tanks of sufficient capacity to hold at least enough acid for four or five nitrations should be placed in a wooden house upon a level at least 6 or 7 feet above the nitrating house. In this house also should be a smaller lead tank, holding, when filled to a certain mark, just enough of the mixed acids for one nitration. The object of this tank is, that as soon as the man in charge knows that the last nitration is finished, he refills this smaller tank (which contains just enough of the mixed acids), and allows its contents to flow down into the nitrating house and into the nitrator, ready for the next nitration. The nitration is usually conducted in a vessel constructed of lead, some 4 feet wide at the bottom, and rather less at the top, and about 4 feet or so high. The size, of course, depends upon the volume of the charge it is intended to nitrate at one operation, but it is always better that the tank should be only two-thirds full. A good charge is 16 cwt. of the mixed acids, in the proportion of three to five; that is, 6 cwt. of nitric acid, and 10 cwt. of sulphuric acid, and 247 lbs. of glycerine.

Upon reference to the equation showing the formation of nitro-glycerine, it will be seen that for every 1 lb. of glycerine 2.47 lbs. of nitro- glycerine should be furnished,[A] but in practice the yield is only a little over 2 lbs., the loss being accounted for by the unavoidable formation of some of the lower nitrate of glycerine (the mononitrate), which afterward dissolves in the washing waters. The lead tank (Fig. 5) is generally cased in woodwork, with a platform in front for the man in charge of the nitrating to stand upon, and whence to work the various taps. The top of the tank is closed in with a dome of lead, in which is a small glass window, through which the progress of the nitrating operation can be watched. From the top of this dome is a tube of lead which is carried up through the roof of the building. It serves as a chimney to carry off the acid fumes which are given off during the nitration. The interior of this tank contains at least three concentric spirals of at least 1-inch lead pipe, through which water can be made to flow during the *whole* operation of nitrating. Another lead pipe is carried through the dome of the tank, as far as the bottom, where it is bent round in the form of a circle. Through this pipe, which is pierced with small holes, about 1 inch apart, compressed air is forced at a pressure of about 60 lbs. in order to keep the liquids in a state of constant agitation during the whole period of nitration. There must also be a rather wide pipe, of say 2 inches internal diameter, carried through the dome of the tank, which will serve to carry the mixed acid to be used in the operation into the tank. There is still another pipe to go through the dome, viz., one to carry the glycerine into the tank. This need not be a large bore pipe, as the glycerine is generally added to the mixed acids in a thin stream (an injector is often used).

[Footnote A: Thus if 92 lbs. glycerine give 227 lbs. nitro-glycerine, $(277 \times 1)/92 = 2.47$ lbs.]

[Illustration: FIG. 5.—TOP OF NITRATOR. *A*, Fume Pipe; *B*, Water Pipes for Cooling; *C*, Acid Mixture Pipe; *E*, Compressed Air; *G*, Glycerine Pipe and Funnel; *T*, Thermometer; *W*, Window.]

Before the apparatus is ready for use, it requires to have two thermometers fixed, one long one to reach to the bottom of the tank, and one short one just long enough to dip under the surface of the acids. When the tank contains its charge, the former gives the temperature of the bottom, and the latter of the top of the mixture. The glycerine should be contained in a small cistern, fixed in some convenient spot upon the wall of the nitrating house, and should have a pipe let in flush with the bottom, and going through the dome of the nitrating apparatus. It must of course be provided with a tap or stop-cock, which should be placed just above the point where the pipe goes through the lead dome.

Some method of measuring the quantity of glycerine used must be adopted. A gauge-tube graduated in inches is a very good plan, but it is essential that the graduations should be clearly visible to the operator upon the platform in front of the apparatus. A large tap made of earthenware (and covered with lead) is fixed in the side of the nitrating tank just above the bottom, to run off the charge after nitration. This should be so arranged that the charge may be at option run down the conduit to the next house or discharged into a drowning tank, which may sometimes be necessary in cases of decomposition. The drowning tank is generally some 3 or 4 yards long and several feet deep, lined with cement, and placed close outside the building.

The apparatus having received a charge of mixed acids, the water is started running through the pipes coiled inside the tank, and a slight pressure of compressed air is turned on,[A] to mix the acids up well before starting. The nitration should not be commenced until the two thermometers register a temperature of 18° C. The glycerine tap is then partially opened, and the glycerine slowly admitted, and the compressed air turned on full, until the contents of the apparatus are in a state of very brisk agitation. A pressure of about 40 lbs. is about the minimum (if 247 lbs. of glycerine and 16 cwt. of acids are in the tank). If the glycerine tube is fitted with an injector, it may be turned on almost at once. The nitration will take about thirty minutes to complete, but the compressed air and water should be kept on for an additional ten minutes after this, to give time for all the glycerine to nitrate. The temperature should be kept as low as possible (not above 18° C.).

[Footnote A: At the Halton Factory, Germany, cylinders of compressed carbon dioxide are connected with the air pipes so that in the event of a failure of the air supply the stirring can be continued with this gas if necessary.]

The chief points to attend to during the progress of the nitration are—

1. The temperature registered by the two thermometers.

2. The colour of the nitrous fumes given off (as seen through the little window in the dome of the apparatus).

3. The pressure of the compressed air as seen from a gauge fixed upon the air pipe just before it enters the apparatus.

4. The gauge showing the quantity of glycerine used. The temperature, as shown by either of the two thermometers, should not be at any time higher than 25° C.

If it rises much above this point, the glycerine should be at once shut off, and the pressure of air increased for some few minutes until the temperature falls, and no more red fumes are given off.

The nitration being finished, the large earthenware tap at the bottom of the tank is opened, and the charge allowed to flow away down the conduit to the next building, i.e., to the separator.

The nitrating house is best built of wood, and should have a close-boarded floor, which should be kept scrupulously clean, and free from grit and sand. A wooden pail and a sponge should be kept in the house in order that the workman may at once clean up any mess that may be made, and a small broom should be handy, in order that any sand, &c., may be at once removed. It is a good plan for the nitrator to keep a book in which he records the time of starting each nitration, the temperature at starting and at the finish, the time occupied, and the date and number of the charge, as this enables the foreman of the danger area at any time to see how many charges have been nitrated, and gives him other useful information conducive to safe working. Edward Liebert has devised an improvement in the treatment of nitro-glycerine. He adds ammonium sulphate or ammonium nitrate to the mixed acids during the operation of nitrating, which he claims destroys the nitrous acid formed according to the equation—

$$(NH_4)_2SO_4 + 2HNO_3 = H_2SO_4 + 2N_2 + 4H_2O.$$

I am not aware that this modification of the process of nitration is in use at the present time.

The newly made charge of nitro-glycerine, upon leaving the nitrating house, flows away down the conduit, either made of rubber pipes, or better still, of woodwork, lined with lead and covered with lids made of wood (in short lengths), in order that by lifting them at any point the condition of the conduit can be examined, as this is of the greatest importance, and the conduit requires to be frequently washed out and the sulphate of lead removed. This sulphate always contains nitro-glycerine, and should therefore be burnt in some spot far removed from any danger building or magazine, as it frequently explodes with considerable violence.

[Illustration: FIG. 6.—SMALL NITRATOR. N, Tap for Discharging; P,
Water Pipes; T, Thermometer; W, Windows; P', Glycerine Pipe.]

In works where the manufacture of nitro-glycerine is of secondary importance, and some explosive containing only perhaps 10 per cent. of nitroglycerine is manufactured, and where 50 or 100 lbs. of glycerine are nitrated at one time, a very much smaller nitrating apparatus than the one that has been already described will be probably all that is required. In this case the form of apparatus shown in Fig. 6 will be found very satisfactory. It should be made of stout lead (all lead used for tanks, &c., must be "chemical lead"), and may be made to hold 50 or 100 lbs. as found most convenient. This nitrator can very well be placed in the same house as the separator;

in fact, where such a small quantity of nitro- glycerine is required, the whole series of operations, nitrating, separation, and washing, &c., may very well be performed in the same building. It will of course be necessary to place the nitrator on a higher level than the separator, but this can easily be done by having platforms of different heights, the nitration being performed upon the highest. The construction of this nitrator is essentially the same as in the larger one, the shape only being somewhat different. Two water coils will probably be enough, and one thermometer. It will not be necessary to cover this form in with woodwork.

~The Nathan Nitrator.~[A] — This nitrator is the patent of Lt. Col. F.L. Nathan and Messrs J.M. Thomson and W. Rintoul of Waltham Abbey, and will probably before long entirely supersede all the other forms of nitrator on account of its efficiency and economy of working. With this nitrator it is possible to obtain from 2.21 to 2.22 parts of nitro-glycerine from every 1 part of glycerine. The apparatus is so arranged that the nitration of the glycerine, the separation of nitro-glycerine produced, as well as the operation of "after-separation," are carried out in one vessel. The usual nitrating vessel is provided with an acid inlet pipe at the bottom, and a glass separation cylinder with a lateral exit or overflow pipe at the top. This cylinder is covered by a glass hood or bell jar during nitration to direct the escaping air and fumes into a fume pipe where the flow of the latter may be assisted by an air injector. The lateral pipe in the separation cylinder is in connection with a funnel leading to the prewash tank. The drawing (Fig. 7) shows a vertical section of the apparatus; a is the nitrating vessel of usual construction, having at the bottom an acid inlet pipe with three branches, one leading to the de-nitrating plant, c leading to the drowning tank, and d, which extends upwards and has two branches, e leading to the nitrating acids tank, and f to the waste acid tank. On the sloped bottom of the nitrating vessel a lies a coil g of perforated pipe for blowing air, and there are in the vessel several coils h, three shown in the drawing, for circulation of cooling water. At the top of the vessel there is a glass cylinder i, having a lateral outlet j directed into the funnel mouth of a pipe k leading to the prewash tank. Over the cylinder i is a glass globe l, into which opens a pipe m for leading off fumes which may be promoted by a compressed air jet from a pipe r operating as an injector. Into an opening of the glass dome l is inserted a vessel n, which is connected by a flexible pipe p to the glycerine tank, and from the bottom of n, which is perforated and covered with a disc perforated with holes registering with those through the bottom, this disc being connected by a stem with a knob q by which it can be turned so as to throttle or cut off passage of glycerine through the bottom. s is a thermometer for indicating the temperature of the contents of the vessel.

[Footnote A: Eng. Pat. 15,983, August 1901.]

[Illustration: FIG. 7.—NATHAN'S NITRATOR FOR NITRO-GLYCERINE. (*a*) Nitrating Vessel; (*b*) to Separating Vessel; (*c*) to Drowning Tank; (*e*) Nitrating Acids enter (*f*) to the Waste Acids; (*g*) Coils for Compressed Air; (*h*) Pipes for Cooling Water; (*i*) Glass Cylinder; (*j*) Outlet to *k*; (*k*) leading to Prewash Tank; (*l*) Glass Dome; (*m*) Pipe to lead off for Escape of Fumes; (*n*) Vessel; (*p*) Pipe conveying Glycerine; (*q*) Knob to turn off Glycerine; (*r*) Compressed Air Jet; (*s*) Thermometer.]

In operating with this apparatus the nitrating acid is introduced into the nitrating vessel by opening the cock of the pipe e. The glycerine is then run in by introducing n and opening the valve at its bottom, the contents of the vessel being agitated by air blown through the perforations of the pipe g. When the glycerine is all nitrated and the temperature has slightly fallen, the circulation of the water through the coils h and the air-stirring are stopped, and the glycerine supply vessel n is removed. The nitro-glycerine as it separates from the acids is raised by introducing by the pipe f waste acid from a previous charge, this displacing the nitro-glycerine upwards and causing it to flow by the outlet, j and pipe k to the prewash tank. When nearly all the nitro-glycerine has been separated in this manner the acids in the apparatus may be run off by the pipe b to an after separating vessel for further settling, thus leaving the apparatus free for another nitration, or the nitrating vessel itself may be used as an after separating bottle displacing the nitroglycerine with waste acid as it rises to the top, or skimming off in the usual manner. When the separation of the nitro- glycerine is complete the waste acid is run off and denitrated as usual, a portion of it being reserved for the displacement of the nitro-glycerine in a subsequent operation.

In a further patent (Eng. Pat. 3,020, 1903) the authors propose with the object of preventing the formation and separation of nitro-glycerine in the waste acids, after the nitro-glycerine initially formed in the nitrating vessel has been separated and removed, to add a small quantity of water to the waste acids; this is carried out as follows. A relatively small quantity of water is added, and this prevents all further separation of nitro-glycerine, and at the same time the strength of the waste acids is so slightly reduced that their separation and re-concentration are not affected. "After-separation" is thus done away with, and the nitro- glycerine plant simplified and its output increased. After nitration separation is commenced at a temperature such that when all the displacing acid has been added, and the separation of the nitro-glycerine is complete, the temperature of the contents of the nitrating vessel shall not be lower than 15° C. A sufficient quantity of the displacing acid is then run off through the waste-acid cock to allow of the remaining acids being air-stirred without splashing over the top. A small quantity of water, from 2 to 3 per cent. according to strength of acid; if waste consists of sulphuric acid (monohydrate), 62 per cent.; nitric acid (anhydrous), 33 per cent. and water 5 per cent.; temperature 15° C., then 2 per cent. of water is added; if waste acids contain less than 4 per cent. of water of temperature lower than 15° C., from 3 to 5 per cent. of water may have to be added. The water is added slowly through the separator cylinder, and the contents of the nitrator air-stirred, but not cooled, the temperature being allowed to rise slowly and regularly as the water is added—usually about 3° C. for each per cent. of water added. When air-agitation has been stopped, the acids are kept at rest for a short time, in order to allow of any small quantity of initially formed nitro-glycerine adhering to the coils and sides of the vessel rising to the top. When this has been separated by displacement, the acids are ready for denitration, or can be safely stored without further precaution.

~Separation.~—The nitro-glycerine, together with the mixed acids, flows from the nitrating house to the separating house, which must be on a lower level than the former. The separating house contains a large lead-lined tank, closed in at the top with a wooden lid, into which a lead pipe of large bore is fixed, and which is carried up through the roof of the building, and acts as a chimney to carry off any fumes. A little glass window should be fixed in this pipe in order that the colour of the escaping fumes may be seen. The conduit conveying the nitro-glycerine enters the building close under the roof, and discharges its contents into the tank through the pipe G (Fig. 8). The tank is only about two-thirds filled by the charge. There is in the side of the tank a small window of thick plate glass, which enables the workman to see the level of the charge, and also to observe the progress of the separation, which will take from thirty minutes to one hour.

The tank should be in connection with a drowning tank, as the charge sometimes gets very dangerous in this building. It must also be connected by a conduit with the filter house, and also to the secondary separator by another conduit. The tank should also be fitted with a compressed air pipe, bent in the form of a loop. It should lie upon the bottom of the vat. The object of this is to mix up the charge in case it should get too hot through decomposition. A thermometer should of course be fixed in the lid of the tank, and its bulb should reach down to the middle of the nitro-glycerine (which rests upon the surface of the mixed acids, the specific gravity of the nitro-glycerine being 1.6, and that of the waste acids 1.7; the composition of the acids is now 11 per cent. HNO_{3}, 67 per cent. $H_{2}SO_{4}$, and 22 per cent. water), and the temperature carefully watched.

[Illustration: FIG. 8.—SEPARATOR. *A*, Compressed Air Pipes; *G*, Nitro-
glycerine enters from Nitrator; *N*, Nitro-glycerine to *P*; *L*, Lantern Window; *W*, Window in Side; *S*, Waste Acids to Secondary Separator; *T*,
Tap to remove last traces of Nitro-glycerine; *P*, Lead Washing Tank; *A*,
Compressed Air; *W*, Water Pipe; *N*, Nitro-glycerine from Separator.]

If nothing unusual occurs, and it has not been necessary to bring the compressed air into use, and so disturb the process of separation, the waste acids may be run away from beneath the nitro-glycerine, and allowed to flow away to the secondary separator, where any further quantity of nitro-glycerine that they contain separates out after resting for some days. The nitro-glycerine itself is run into a smaller tank in the same house, where it is washed three or four times with its own bulk of water, containing about 3 lbs. of carbonate of soda to neutralise the remaining acid. This smaller tank should contain a lead pipe, pierced and coiled upon the bottom, through which compressed air may be passed, in order to stir up the charge with the water and soda. After this preliminary washing, the nitro-glycerine is drawn off into indiarubber buckets, and poured down the conduit to the filter house. The wash waters may be sent down a conduit to another building, in order to allow the small quantity of nitro-glycerine that has been retained in the water as minute globules to settle, if thought worth the trouble of saving. This, of course, will depend upon the usual out-turn of nitro-glycerine in a day, and the general scale of operations.

[Illustration: FIG. 9.—FILTERING AND WASHING PLANT. *W*, Lead Washing
Tank; *WP*, Water Pipe; *L*, Lid; *S*, Nitro-glycerine from Separator; *A, B, C*, Filtering Tanks; *B2*, Indiarubber Bucket.]

~Filtering and Washing.~—The filter house (Fig. 9), which must of course be again on a somewhat lower level than the separating house, must be a considerably larger building than either the nitrating or separating houses, as it is always necessary to be washing some five or six charges at the same time. Upon the arrival of the nitro-glycerine at this house, it first flows into a lead-lined wooden tank (W), containing a compressed air pipe, just like the one in the small tank in the separating house. This tank is half filled with water, and the compressed air is turned on from half to a quarter of an hour after the introduction of the charge. The water is then drawn off, and fresh water added. Four or five washings are generally necessary. The nitro-glycerine is then run into the next tank (A), the top of which is on a level with the bottom of the first one. Across the top of this tank is stretched a frame of flannel, through which the nitroglycerine has to filter. This removes any solid matters, such as dirt or scum. Upon leaving this tank, it passes through a similar flannel frame across another tank (B), and is finally drawn off by a tap in the bottom of the tank into rubber buckets. The taps in these tanks are best made of vulcanite.

At this stage, a sample should be taken to the laboratory and tested. If the sample will not pass the tests, which is often the case, the charge must be rewashed for one hour, or some other time, according to the judgment of the chemist in charge. In the case of an obstinate charge, it is of much more avail to wash a large number of times with small quantities of water, and for a short time, than to use a lot of water and wash for half an hour. Plenty of compressed air should be used, as the compound nitric ethers which are formed are thus got rid of. As five or six charges are often in this house at one time, it is necessary to have as many tanks arranged in tiers, otherwise one or two refractory charges would stop the nitrating house and the rest of the nitro-glycerine plant. The chief causes of the washed material not passing the heat test are, either that the acids were not clean, or they contained objectionable impurities, or more frequently, the quality of the glycerine used. The glycerine used for making nitro-glycerine should conform to the following tests, some of which, however, are of greater importance than others. The glycerine should—

1. Have minimum specific gravity at 15° C. of 1.261.

2. Should nitrify well.

3. Separation should be sharp within half an hour, without the separation of flocculent matter, nor should any white flocculent matter (due to fatty acids) be formed when the nitrated glycerine is thrown into water and neutralised with carbonate of soda.

4. Should be free from lime and chlorine, and contain only traces of arsenic, sulphuric acid, &c.

5. Should not leave more than 0.25 per cent. of inorganic and organic residue together when evaporated in a platinum dish without ebullition (about 160° C.) or partial decomposition.

6. Silver test fair.

7. The glycerine, when diluted one-half, should give no deposit or separation of fatty acids when nitric peroxide gas is passed through it. (Nos. 1, 2, 3, and 5 are the most essential.)

The white flocculent matter sometimes formed is a very great nuisance, and any sample of glycerol which gives such a precipitate when tried in the laboratory should at once be rejected, as it will give no end of trouble in the separating house, and also in the filter house, and it will be very difficult indeed to make the nitro-glycerine pass the heat test. The out- turn of nitro-glycerine also will be very low. The trouble will show itself chiefly in the separating operation. Very often 2 or 3 inches will rise to the surface or hang about in the nitro-glycerine, and at the point of contact between it and the mixed acids, and will afterwards be very difficult to get rid of by filtration. The material appears to be partly an emulsion of the glycerine, and partly due to fatty acids, and as there appears to be no really satisfactory method of preventing its formation, or of getting rid of it, the better plan is not to use any glycerine for nitrating that has been found by experiment upon the laboratory scale to give this objectionable matter. One of the most useful methods of testing the glycerine, other than nitrating, is to dilute the sample one-half with water, and then to pass a current of nitric peroxide gas through it, when a flocculent precipitate of elaïdic acid (less soluble in glycerine than the original oleic acid) will be formed. Nitrogen peroxide, N_2O_4, is best obtained by heating dry lead nitrate (see Allen, "Commercial Organic Analysis," vol. ii., 301).

When a sample of nitro-glycerine is brought to the laboratory from the filter house, it should first be examined to see that it is not acid.[A] A weak solution of Congo red or methyl orange may be used. If it appears to be decidedly alkaline, it should be poured into a separating funnel, and shaken with a little distilled water. This should be repeated, and the washings (about 400 c.c.) run into a beaker, a drop of Congo red or methyl orange added, and a drop or so of N/2 hydrochloric acid added, when it should give, with two or three drops at most, a blue colour with the Congo red, or pink with the methyl orange, &c. The object of this test is to show that the nitro-glycerine is free from any excess of soda, i.e., that the soda has been properly washed out, otherwise the heat test will show the sample to be better than it is. The heat test must also be applied.

[Footnote A: A. Leroux, *Bul. Soc. Chim. de Bel.*, xix., August 1905, contends that experience does not warrant the assumption that free acid is a source of danger in nitro-glycerine or nitro-cellulose; free alkali, he states, promotes their decomposition.]

Upon leaving the filter house, where it has been washed and filtered, and has satisfactorily passed the heat test, it is drawn off from the lowest tank in indiarubber buckets, and poured down the conduit leading to the precipitating house, where it is allowed to stand for a day, or sometimes longer, in order to allow the little water it still contains to rise to the surface. In order to accomplish this, it is sufficient to allow it to stand in covered-in tanks of a conical form, and about 3 or 4 feet high. In many works it is previously filtered through common salt, which of course absorbs the last traces of water. It is then of a pale yellow colour, and should be quite clear, and can be drawn off by means of a tap (of vulcanite), fixed at the bottom of the tanks, into rubber buckets, and is ready for use in the preparation of dynamite, or any of the various forms of gelatine compounds, smokeless powders, &c., such as cordite, ballistite, and many others.

Mikolajezak (*Chem. Zeit.*, 1904, Rep. 174) states that he has prepared mono- and di-nitro-glycerine, and believes that the latter compound will form a valuable basis for explosives, as it is unfreezable. It is stated to be an odourless, unfreezable oil, less sensitive to percussion, friction, and increase of temperature, and to possess a greater solvent power for collodion-cotton than ordinary nitro-glycerine. It can thus be used for the preparation of explosives of high stability, which will maintain their plastic nature even in winter. The di-nitro-glycerine is a solvent for tri-nitro-glycerine, it can therefore be mixed with this substance, in the various gelatine explosives in order to lower the freezing point.

~The Waste Acids.~ — The waste acids from the separating house, from which the nitro-glycerine has been as completely separated as possible, are run down the conduit to the secondary separator, in order to recover the last traces of nitro-glycerine that they contain. The composition of the waste acids is generally somewhat as follows: — Specific gravity, 1.7075 at 15° C.; sulphuric acid, 67.2 per cent.; nitric acid, 11.05 per cent.; and water, 21.7 per cent., with perhaps as much as 2 per cent. of nitric oxide, and of course varying quantities of nitro-glycerine, which must be separated, as it is impossible to run this liquid away (unless it can be run into the sea) or to recover the acids by distillation as long as it contains this substance. The mixture, therefore, is generally run into large circular lead-lined tanks, covered in, and very much like the nitrating apparatus in construction, that is, they contain worms coiled round inside, to allow of water being run through to keep the mixture cool, and a compressed air pipe, in order to agitate the mixture if necessary. The top also should contain a window, in order to allow of the interior being seen, and should have a leaden chimney to carry off the fumes which may arise from decomposition. It is also useful to have a glass tube of 3 or 4 inches in diameter substituted for about a foot of the lead chimney, in order that the man on duty can at any time see the colour of the fumes arising from the liquid. There should also be two thermometers, one long one reaching to the bottom of the tank, and one to just a few inches below the surface of the liquid.

The nitro-glycerine, of course, collects upon the surface, and can be drawn off by a tap placed at a convenient height for the purpose. The cover of the tank is generally conical, and is joined to a glass cylinder, which is cemented to the top of this lead cover, and also to the lead chimney. In this glass cylinder is a hole into which fits a ground glass stopper, through which the nitro-glycerine can be drawn off. There will probably never be more than an inch of nitro-glycerine at the most, and seldom that. It should be taken to the filter house and treated along with another charge. The acids themselves may either be run to waste, or better treated by some denitration plant. This house probably requires more attention than any other in the danger area, on account of the danger of the decomposition of the small quantities of nitro-glycerine, which, as it is mixed with such a large quantity of acids and water, is very apt to become hot, and decomposition, which sets up in spots where a little globule of nitro-glycerine is floating, surrounded by acids that gradually get hot, gives off nitrous fumes, and perhaps explodes, and thus causes the sudden explosion of the whole. The only way to prevent this is for the workman in charge to look at the thermometers *frequently*, and at the colour of the escaping fumes, and if he should notice a rise of temperature or any appearance of red fumes, to turn on the water and air, and stir up the mixture, when probably the temperature will suddenly fall, and the fumes cease to come off.

The cause of explosions in this building is either the non-attention of the workmen in charge, or the bursting of one of the water pipes, by which means, of course, the water, finding its way into the acids, causes a sudden rise of temperature. If the latter of these two causes should occur, the water should at once be shut off and the air turned on full, but if it is seen that an explosion is likely to occur, the tank should at once be emptied by allowing its contents to run away into a drowning tank placed close outside the house, which should be about 4 feet deep, and some 16 feet long by 6 feet wide; in fact, large enough to hold a considerable quantity of water. But this last course should only be resorted to as a last extremity, as it is extremely troublesome to recover the small quantity of nitro-glycerine from the bottom of this tank, which is generally a bricked and cemented excavation some few yards from the house.

It has been proposed to treat these waste acids, containing nitro-glycerine, in Mr M. Prentice's nitric acid retort. In this case they would be run into the retort, together with nitrate of soda, in a fine stream, and the small quantity of nitro-glycerine, coming into contact with the hot mixture already in the retort, would probably be at once decomposed. This process, although not yet tried, promises to be a success. Several processes have been used for the denitration of these acids.

~Treatment of the Waste Acid from the Manufacture of Nitro-Glycerine and
Gun-Cotton.~ — The composition of these acids is as follows: —

Nitro-glycerine and Gun-cotton
Waste Acid.

Sulphuric acid 70 per cent. 78 per cent.
Nitric acid 10 " 12 "
Water 20 " 10 "

The waste acid from the manufacture of gun-cotton is generally used direct for the manufacture of nitric acid, as it contains a fairly large amount of sulphuric acid, and the small amount of nitro-cellulose which it also generally contains decomposes gradually and without explosion in the retort. Nitric acid may be first distilled off, the resulting sulphuric acid being then added to the equivalent amount of nitrate of soda. Nitric acid is then distilled over and condensed in the usual way. Very often, however, the waste acid is added direct to the charge of nitrate without previously eliminating the nitric acid. The treatment of the waste acid from the manufacture of nitro-glycerine is somewhat different. The small amount of nitro-glycerine in this acid must always be eliminated. This is effected either by allowing the waste acid to stand for at least twenty-four hours in a big vessel with a conical top, where all the nitro-glycerine which will have separated to the surface is removed by skimming; or, better still, the "watering down process" of Col. Nathan may be employed. In Nathan's nitrator every existing trace of nitro-glycerine is separated from the acids in a few hours after the nitration, and any further formation of nitro-glycerine is prevented by adding about 2 per cent. of water to the waste acids, which are kept agitated during the addition. The waste acid, now free from nitro-glycerine, but which may still contain organic matter, is denitrated by bringing it into contact with a jet of steam. The waste acid is passed in a small stream down through a tower of acid-resisting stoneware (volvic stone), which is closely packed with earthenware, and at the bottom of which is the steam jet. Decomposition proceeds as the acid meets the steam, nitric and nitrous acids are disengaged and are passed out at the top of the tower through a pipe to a series of condensers and towers, where the nitric acid is collected. The nitrous acid may be converted into nitric acid by introducing a hot compressed air jet into the gases before they pass into the condensers. Weak sulphuric acid of sp. gr. 1.6 collects in a saucer in which the tower stands, and is then passed through a cooling worm. The weak sulphuric acid, now entirely free from nitric and nitrous acids, may be concentrated to sp. gr. 1.842 and 96 per cent. H_2SO_4 by any of the well-known processes, e.g., Kessler, Webb, Benker, Delplace, &c., and it may be used again in the manufacture of nitro-glycerine or gun-cotton.

Two points in the manufacture of nitro-glycerine are of the greatest importance, viz., the purity of the glycerine used, and the strength and purity of the acids used in the nitration. With regard to the first of these, great care should be taken, and a complete analysis and thorough examination, including a preliminary experimental nitration, should always be instituted. As regards the second, the sulphuric acid should not only be strong (96 per cent.), but as free from impurities as possible. With the nitric acid, which is generally made at the explosive works where it is used, care must be taken that it is as strong as possible (97 per cent. and upwards). This can easily be obtained if the plant designed by Mr Oscar Guttmann[A] is used. Having worked Mr Guttmann's plant for some time, I can testify as to its value and efficiency.

[Footnote A: "The Manufacture of Nitric Acid," *Jour. Soc. Chem. Ind.*, March 1893.]

Another form of nitric acid plant, which promises to be of considerable service to the manufacturer of nitric acid for the purpose of nitrating, is the invention of the late Mr Manning Prentice, of Stowmarket. Through the kindness of Mr Prentice, I visited his works to see the plant in operation. It consists of a still, divided into compartments or chambers in such a manner that the fluid may pass continuously from one to the other. The nitric acid being continuously separated by distillation, the contents of each division vary — the first containing the full proportion of nitric acid, and each succeeding one less of the nitric acid, until from the overflow of the last one the bisulphate of soda flows away without any nitric acid. The nitrate of soda is placed in weighed quantities in the hopper, whence it passes to the feeder. The feeder is a miniature horizontal pug-mill, which receives the streams of sulphuric acid and of nitrate, and after thoroughly mixing them, delivers them into the still, where, under the influence of heat, they rapidly become a homogeneous liquid, from which nitric acid continuously distils.

Mr Prentice says: "I may point out that while the ordinary process of making nitric acid is one of fractional distillation by time, mine is fractional distillation by space." "Instead of the operation being always at the same point of space, but differing by the successive points of time, I arrange for the differences to take place at different points of space, and these differences exist at one and the same points of time." It is possible with this plant to produce the full product of nitric acid of a gravity of 1.500, or to obtain the acid of varying strengths from the different still-heads. One of these stills, capable of producing about 4 tons of nitric acid per week, weighs less than 2 tons. It is claimed that there is by their use a saving of more than two-thirds in fuel, and four- fifths in condensing plant. Further particulars and illustrations will be found in Mr Prentice's paper (*Journal of the Society of Chemical Industry*, 1894, p. 323).

CHAPTER III.

NITRO-CELLULOSE, &c.

Cellulose Properties — Discovery of Gun-Cotton — Properties of Gun-Cotton —
Varieties of Soluble and Insoluble Gun-Cottons — Manufacture of Gun-Cotton —
Dipping and Steeping — Whirling out the Acid — Washing — Boiling — Pulping —
Compressing — The Waltham Abbey Process — Le Bouchet Process — Granulation of
Gun-Cotton — Collodion-Cotton — Manufacture — Acid Mixture used — Cotton used,
&c. — Nitrated Gun-Cotton — Tonite — Dangers in Manufacture of Gun-Cotton —
Trench's Fire-Extinguishing Compound — Uses of Collodion-Cotton — Celluloid —
Manufacture, &c. — Nitro-Starch, Nitro-Jute, and Nitro-Mannite.

~The Nitro-Celluloses.~ — The substance known as cellulose forms the groundwork of vegetable tissues. The cellulose of the woody parts of plants was at one time supposed to be a distinct body, and was called lignine, but they are now regarded as identical. The formula of cellulose is $(C_6H_{10}O_6)\{X\}$, *and it is generally assumed that the molecular formula must be represented by a multiple of the empirical formula,* $C_{12}H_{20}O_{10}$ being often regarded as the minimum. The assumption is based on the existence of a pentanitrate and the insoluble and colloidal nature of cellulose. Green (*Zeit. Farb. Text. Ind.*, 1904, 3, 97) considers these reasons insufficient, and prefers to employ the single formula $C_6H_{10}O_5$. Cellulose can be extracted in the pure state, from young and tender portions of plants by first crushing them, to rupture the cells, and then extracting with dilute hydrochloric acid, water, alcohol, and ether in succession, until none of these solvents remove anything more. Fine paper or cotton wool yield very nearly pure cellulose by similar treatment.

Cellulose is a colourless, transparent mass, absolutely insoluble in water, alcohol, or ether. It is, however, soluble in a solution of cuprammoniac solution, prepared from basic carbonate or hydrate of copper and aqueous ammonia. The specific gravity of cellulose is 1.25 to 1.45. According to Schulze, its elementary composition is expressed by the percentage numbers: —

Carbon 44.0 per cent. 44.2 per cent.
Hydrogen 6.3 " 6.4 "
Oxygen 49.7 " 49.4 "

These numbers represent the composition of the ash free cellulose. Nearly all forms of cellulose, however, contain a small proportion of mineral matters, and the union of these with the organic portion of the fibre or tissue is of such a nature that the ash left on ignition preserves the form of the original. "It is only in the growing point of certain young shoots that the cellulose tissue is free from mineral constituents" (Hofmeister).

Cellulose is a very inert body. Cold concentrated sulphuric acid causes it to swell up, and finally dissolves it, forming a viscous solution. Hydrochloric acid has little or no action, but nitric acid has, and forms a series of bodies known as nitrates or nitro-celluloses. Cellulose has some of the properties of alcohols, among them the power of forming ethereal salts with acids. When cellulose in any form, such as cotton, is brought into contact with strong nitric acid at a low temperature, a nitrate or nitro product, containing nitryl, or the NO_2 group, is produced. The more or less complete replacement of the hydroxylic hydrogen by NO_2 groups depends partly on the concentration of the nitric acid used, partly on the duration of the action. If the most concentrated nitric and sulphuric acids are employed, and the action allowed to proceed for some considerable time, the highest nitrate, known as hexa-nitro- cellulose or guncotton, $C_{12}H_{14}O_4(O.NO_2)_6$, will be formed; but with weaker acids, and a shorter exposure to their action, the tetra and penta and lower nitrates will be formed.[A]

[Footnote A: The paper by Prof. Lunge, *Jour. Amer. Chem. Soc.*, 1901, 23[8], 527-579, contains valuable information on this subject.]

Besides the nitrate, A. Luck[A] has proposed to use other esters of cellulose, such as the acetate, benzoate, or butyrate. It is found that cellulose acetate forms with nitro-glycerine a gelatinous body without requiring the addition of a solvent. A sporting powder is proposed composed of 75 parts of cellulose nitrate (13 per cent. N.) mixed with 13 parts of cellulose acetate.

[Footnote A: Eng. Pat. 24,662, 22nd November 1898.]

The discovery of gun-cotton is generally attributed to Schönbein (1846), but Braconnot (in 1832) had previously nitrated starch, and six years later Pelouse prepared nitro-cotton and various other nitro bodies, and Dumas nitrated paper, but Schönbein was apparently the first chemist to use a mixture of strong nitric and sulphuric acids. Many chemists, such as Piobert in France, Morin in Russia, and Abel in England, studied the subject; but it was in Austria, under the auspices of Baron Von Lenk, that the greatest progress was made. Lenk used cotton in the form of yarn, made up into hanks, which he first washed in a solution of potash, and then with water, and after drying dipped them in the acids. The acid mixture used consisted of 3 parts by weight of sulphuric to 1 part of nitric acid, and were prepared some time before use. The cotton was dipped one skein at a time, stirred for a few minutes, pressed out, steeped, and excess of acid removed by washing with water, then with dilute potash, and finally with water. Von Lenk's process was used in England at Faversham (Messrs Hall's Works), but was given up on account of an explosion (1847).

Sir Frederick Abel, working at Stowmarket and Waltham Abbey, introduced several very important improvements into the process, the chief among these being pulping. Having traced the cause of its instability to the presence of substances caused by the action of the nitric acid on the resinous or fatty substances contained in the cotton fibre, he succeeded in eliminating them, by boiling the nitro-cotton in water, and by a thorough washing, after pulping the cotton in poachers.

Although gun-cottons are generally spoken of as nitro-celluloses, they are more correctly described as cellulose nitrates, for unlike nitro bodies of other series, they do not yield, or have not yet done so, amido bodies, on reduction with nascent hydrogen.[A] The equation of the formation of gun-cotton is as follows:—

$2(C_6H_{10}O_5) + 6HNO_3 = C_{12}H_{14}O_4(NO_3)_6 + 6OH_2$. Cellulose. Nitric Acid. Gun-Cotton. Water.

The sulphuric acid used does not take part in the reaction, but its presence is absolutely essential to combine with the water set free, and thus to prevent the weakening of the nitric acid. The acid mixture used at Waltham Abbey consists of 3 parts by weight of sulphuric acid of 1.84 specific gravity, and 1 part of nitric acid of 1.52 specific gravity. The same mixture is also used at Stowmarket (the New Explosive Company's Works). The use of weaker acids results in the formation of collodion- cotton and the lower nitrates generally.

[Footnote A: "Cellulose," by Cross and Bevan, ed. by W.R. Hodgkinson, p. 9.]

The nitrate which goes under the name of gun-cotton is generally supposed to be the hexa-nitrate, and to contain 14.14 per cent. of nitrogen; but a higher percentage than 13.7 has not been obtained from any sample. It is almost impossible (at any rate upon the manufacturing scale) to make pure hexa-nitro-cellulose or gun-cotton; it is certain to contain several per cents. of the soluble forms, i.e., lower nitrates. It often contains as much as 15 or 16 per cent., and only from 13.07[A] to 13.6 per cent. of nitrogen.

[Footnote A: Mr J.J. Sayers, in evidence before the court in the "Cordite Case," says he found 15.2 and 16.1 per cent. soluble cotton, and 13.07 and 13.08 per cent. nitrogen in two samples of Waltham Abbey gun-cotton.]

A whole series of nitrates of cellulose are supposed to exist, the highest member being the hexa-nitrate, and the lowest the mono-nitrate. Gun-cotton was at one time regarded as the tri-nitrate, and collodion-cotton as the di-nitrate and mono-nitrate, their respective formula being given as follows:—

Mono-nitro-cellulose $C_6H_9(NO_2)O_5$ = 6.763 per cent. nitrogen.

Di-nitro-cellulose $C_6H_8(NO_2)_2O_5$ = 11.11 " "
Tri-nitro-cellulose $C_6H_7(NO_2)_3O_5$ = 14.14 " "

But gun-cotton is now regarded as the hexa-nitrate, and collodion-cotton as a mixture of all the other nitrates. In fact, chemists are now more inclined to divide nitro-cellulose into the soluble and insoluble forms, the reason being that it is quite easy to make a nitro-cellulose entirely soluble in a mixture of ether-alcohol, and yet containing as high a percentage of nitrogen as 12.6; whereas the di-nitrate[A] should theoretically only contain 11.11 per cent. On the other hand, it is not possible to make gun-cotton with a higher percentage of nitrogen than about 13.7, even when it does not contain any nitro-cotton that is soluble in ether-alcohol.[B] The fact is that it is not at present possible to make a nitro-cellulose which shall be either entirely soluble or entirely insoluble, or which will contain the theoretical content of nitrogen to suit any of the above formulæ for the cellulose nitrates. Prof. G. Lunge gives the following list of nitration products of cellulose:—

[Footnote A: The penta-nitrate $C_{12}H_{15}O_5(NO_3)_5$ = 12.75 per cent. nitrogen.]

[Footnote B: In the Cordite Trial (1894) Sir F.A. Abel said, "Before 1888 there was a broad distinction between soluble and insoluble nitro- cellulose, collodion-cotton being soluble (in ether-alcohol) and gun-cotton insoluble." Sir H.E. Roscoe, "That he had been unable to make a nitro-cotton with a higher nitrogen content than 13.7." And Professor G. Lunge said, "Gun-cotton always contained soluble cotton, and *vice versa*." These opinions were also generally confirmed by Sir E. Frankland, Sir W. Crookes, Dr Armstrong, and others.]

Dodeca-nitro-cellulose $C_{24}H_{28}O_{20}(NO_2)_{12}$ = *14.16 per cent.*
 nitrogen. (= *old tri-nitro-cellulose*)
Endeca-nitro-cellulose $C_{24}H_{29}O_{20}(NO_2)_{11}$ = *13.50 per cent.*
 nitrogen.
Deca-nitro-cellulose $C_{24}H_{30}O_{20}(NO_2)_{10}$ = *12.78 per cent.*
 nitrogen.
Ennea-nitro-cellulose $C_{24}H_{31}O_{20}(NO_2)_9$ = *11.98 per cent.*

nitrogen.

Octo-nitro-cellulose $C_{24}H_{32}O_{20}(NO_2)_8$ = 11.13 per cent. nitrogen. (= old di-nitro-cellulose)

Hepta-nitro-cellulose $C_{24}H_{33}O_{20}(NO_2)_7$ = 10.19 per cent. nitrogen.

Hexa-nitro-cellulose $C_{24}H_{34}O_{20}(NO_2)_6$ = 9.17 per cent. nitrogen.

Penta-nitro-cellulose $C_{24}H_{35}O_{20}(NO_2)_5$ = 8.04 per cent. nitrogen.

Tetra-nitro-cellulose $C_{24}H_{36}O_{20}(NO_2)_4$ = 6.77 per cent. nitrogen. (= old mono-nitro-cellulose)

It is not unlikely that a long series of nitrates exists. It is at any rate certain that whatever strength of acids may be used, and whatever temperature or other conditions may be present during the nitration, that the product formed always consists of a mixture of the soluble and insoluble nitro-cellulose.

Theoretically 100 parts of cotton by weight should produce 218.4 parts of gun-cotton, but in practice the yield is a good deal less, both in the case of gun-cotton or collodion-cotton. In speaking of soluble and insoluble nitro-cellulose, it is their behaviour, when treated with a solution consisting of 2 parts ether and 1 of alcohol, that is referred to. There is, however, another very important difference, and that is their different solubility in nitro-glycerine. The lower nitrates or soluble form is soluble in nitro-glycerine under the influence of heat, a temperature of about 50° C. being required. At lower temperatures the dissolution is very imperfect indeed; and after the materials have been left in contact for days, the threads of the cotton can still be distinguished. The insoluble form or gun-cotton is entirely *insoluble* in nitro-glycerine. It can, however, be made to dissolve[A] by the aid of acetone or acetic ether. Both or rather all the forms of nitro-cellulose can be dissolved in acetone or acetic ether. They also dissolve in concentrated sulphuric acid, and the penta-nitrate in nitric acid at about 80° or 90° C.

[Footnote A: Or rather to form a transparent jelly.]

The penta-nitrate may be obtained in a pure state by the following process, devised by Eder: — The gun-cotton is dissolved in concentrated nitric acid at 90° C., and reprecipitated by the addition of concentrated sulphuric acid. After cooling to 0° C., and mixing with a larger volume of water, the precipitated nitrate is washed with water, then with alcohol, dissolved in ether-alcohol, and again precipitated with water, when it is obtained pure. This nitrate is soluble in ether-alcohol, and slightly in acetic acid, easily in acetone, acetic ether, and methyl-alcohol, insoluble in alcohol. Strong potash (KOH) solution converts into the di-nitrate $C_{12}H_{18}O_{8}(NO_{3})_{2}$. The hexa-nitrate is not soluble in acetic acid or methyl-alcohol.

The lower nitrates known as the tetra- and tri-nitrates are formed together when cellulose is treated with a mixture of weak acids, and allowed to remain in contact with them for a very short time (twenty minutes). They cannot be separated from one another, as they all dissolve equally in ether-alcohol, acetic ether, acetic acid, methyl-alcohol, acetone, amyl acetate, &c.

As far as the manufacture of explosive bodies is concerned, the two forms of nitro-cellulose used and manufactured are gun-cotton or the hexa- nitrate (once regarded as tri-nitro-cellulose), which is also known as insoluble gun-cotton, and the soluble form of gun-cotton, which is also known as collodion, and consists of a mixture of several of the lower nitrates. It is probable that it chiefly consists, however, of the next highest nitrate to gun-cotton, as the theoretical percentage of nitrogen for this body,. the penta-nitrate, is 12.75 per cent., and analyses of commercial collodion-cotton, entirely soluble in ether-alcohol, often give as high a percentage as 12.6.

We shall only describe the manufacture of the two forms known as soluble and insoluble, and shall refer to them under their better known names of gun-cotton and collodion-cotton. The following would, however, be the formulæ[A] and percentage of nitrogen of the complete series: —

Hexa-nitro-cellulose $C_{12}H_{14}O_{4}(NO_{3})_{6}$ *14.14 per cent. nitrogen.*

Penta-nitro-cellulose $C_{12}H_{15}O_{5}(NO_{3})_{5}$ *12.75 per cent. nitrogen.*

Tetra-nitro-cellulose $C_{12}H_{16}O_{6}(NO_{3})_{4}$ *11.11 per cent. nitrogen.*

Tri-nitro-cellulose $C_{12}H_{17}O_7(NO_3)_3$ 9.13 per cent. nitrogen.
Di-nitro-cellulose $C_{12}H_{18}O_8(NO_3)_2$ 7.65 per cent. nitrogen.
Mono-nitrocellulose $C_{12}H_{19}O_9(NO_3)$ 3.80 per cent. nitrogen.

[Footnote A: Berthelot takes $C_{24}H_{40}O_{20}$ as the formula of cellulose; and M. Vieille regards the highest nitrate as $(C_{24}H_{18}(NO_3H)_{11}O_9)$. *Compt. Rend.*, 1882, p. 132.]

~Properties of Gun-Cotton.~ — The absolute density of gun-cotton is 1.5. When in lumps its apparent density is 0.1; if twisted into thread, 0.25; when subjected, in the form of pulp, to hydraulic pressure, 1.0 to 1.4. Gun-cotton preserves the appearance of the cotton from which it is made. It is, however, harsher to the touch; it is only slightly hygroscopic (dry gun-cotton absorbs 2 per cent. of moisture from the air). It possesses the property of becoming electrified by friction. It is soluble in acetic ether, amyl acetate, and acetone, insoluble in water, alcohol, ether, ether-alcohol, methyl-alcohol, &c. It is very explosive, and is ignited by contact with an ignited body, or by shock, or when it is raised to a temperature of 172° C. It burns with a yellowish flame, almost without smoke, and leaves little or no residue. The volume of the gases formed is large, and consists of carbonic acid, carbonic oxide, nitrogen, and water gas. Compressed gun-cotton when ignited often explodes when previously heated to 100° C.

Gun-cotton kept at 80° to 100° C. decomposes slowly, and sunlight causes it to undergo a slow decomposition. It can, however, be preserved for years without undergoing any alteration. It is very susceptible to explosions by influence. For instance, a torpedo, even placed at a long distance, may explode a line of torpedoes charged with gun-cotton. The velocity of the propagation of the explosion in metallic tubes filled with pulverised gun-cotton has been found to be from 5,000 to 6,000 mms. per second in tin tubes, and 4,000 in leaden tubes (Sebert).

Gun-cotton loosely exposed in the open air burns eight times as quickly as powder (Piobert). A thin disc of gun-cotton may be fired into from a rifle without explosion; but if the thickness of the disc be increased, an explosion may occur. The effect of gun-cotton in mines is very nearly the same as that of dynamite for equal weights. It requires, however, a stronger detonator, and it gives rise to a larger quantity of carbonic oxide gas. Gun-cotton should be neutral to litmus, and should stand the Government heat test—temperature of 150° F. for fifteen minutes (see page 249). In the French Navy gun-cotton is submitted to a heat test of 65° C. (= 149° F.) for eleven minutes. It should contain as small a percentage of soluble nitro-cotton and of non-nitrated cotton as possible.

The products of perfectly detonated gun-cotton may be expressed by the following equation:—

$2C_{12}H_{14}O_4(NO_3)_6 = 18CO + 6CO_2 + 14H_2O + 12N$.

It does not therefore contain sufficient oxygen for the complete combustion of its carbon. It is for this reason that when used for mining purposes a nitrate is generally added to supply this defect (as, for instance, in tonite). It tends also to prevent the evolution of the poisonous gas, carbonic oxide. The success of the various gelatine explosives is due to this fact, viz., that the nitro-glycerine has an excess of oxygen, and the nitro-cotton too little, and thus the two explosives help one another.

In practice the gases resulting from the explosion of gun-cotton are— Carbonic oxide, 28.55; carbonic acid, 19.11; marsh gas (CH_4), 11.17; nitric oxide, 8.83; nitrogen, 8.56; water vapour, 21.93 per cent. The late Mr E.O. Brown, of Woolwich Arsenal, discovered that perfectly wet and uninflammable compressed gun-cotton could be easily detonated by the detonation of a priming charge of the dry material in contact with it. This rendered the use of gun-cotton very much safer for use as a military or mining explosive.

As a mining explosive, however, gun-cotton is now chiefly used under the form of tonite, which is a mixture of half gun-cotton and half barium nitrate. This material is sometimes spoken of as "nitrated gun-cotton." The weight of gun-cotton required to produce an equal effect either in heavy ordnance or in small arms is to the weight of gunpowder in the proportion of 1 to 3, i.e., an equal weight of gun-cotton would produce three times the effect of gun-powder. Its rapidity of combustion, however, requires to be modified for use in firearms. Hence the lower nitrates are generally used, or such compounds as nitro-lignose, nitrated wood, &c., are used.

The initial pressure produced by the explosion of gun-cotton is very large, equal to 18,135 atmospheres, and 8,740 kilogrammes per square centimetre for 1 kilo., the heat liberated being 1,075 calories (water liquid), or 997.7 cals. (water gaseous), but the quantity of heat liberated changes with the equation of decomposition. According to Berthelot,[A] the heat of formation of collodion-cotton is 696 cals. for 1,053 grms., or 661 cals. for 1 kilo. The heat liberated in the total combustion of gun-cotton by free oxygen at constant pressure is 2,633 cals. for 1,143 grms., or for 1 kilo. gun-cotton 2,302 cals. (water liquid), or 2,177 cals. (water gaseous). The heat of decomposition of gun- cotton in a closed vessel, found by experiment at a low density of charge (0.023), amounts to 1,071 cals. for 1 kilo. of the substance, dry and free from ash. To obtain the maximum effect of gun-cotton it must be used in a compressed state, for the initial pressures are thereby increased. Wet gun-cotton s much less sensitive to shock than dry. Paraffin also reduces its liability to explode, so also does camphor.

[Footnote A: "Explosives and their Power," trans. by Hake and M'Nab.]

The substance known as celluloid, a variety of nitro-cellulose nearly corresponding to the formula $C_{24}H_{24}(NO_3H)_{8}O_{12}$, to which camphor and various inert substances are added, so as to render it non-sensitive to shock, may be worked with tools, and turned in the lathe in the same manner as ivory, instead of which material celluloid is now largely used for such articles as knife handles, combs, &c. Celluloid is very plastic when heated towards 150° C., and tends to become very sensitive to shock, and in large quantities might become explosive during a fire, owing to the general heating of the mass, and the consequent evaporation of the camphor. When kept in the air bath at 135° C., celluloid decomposes quickly. In an experiment (made by M. Berthelot) in a closed vessel at 135° C., and the density of the charge being 0.4, it ended in exploding, developing a pressure of

3,000 kilos. A large package of celluloid combs also exploded in the guard's van on one of the German railways a few years ago. Although it is not an explosive under ordinary circumstances, or even with a powerful detonator, considerable care should be exercised in its manufacture.

~The Manufacture of Gun-Cotton.~ — The method used for the manufacture of gun-cotton is that of Abel (Spec. No. 1102, 20. 4. 65). It was worked out chiefly at Stowmarket[A] and Waltham Abbey,[B] but has in the course of time undergone several alterations. These modifications have taken place, however, chiefly upon the Continent, and relate more to the apparatus and machinery used than to any alteration in the process itself. The form of cellulose used is cotton-waste,[C] which consists of the clippings and waste material from cotton mills. After it has been cleaned and purified from grease, oil, and other fatty substances by treatment with alkaline solutions, it is carefully picked over, and every piece of coloured cotton rag or string carefully removed. The next operation to which it is submitted has for its object the opening up of the material. For this purpose it is put through a carding machine, and afterwards through a cutting machine, whereby it is reduced to a state suitable for its subsequent treatment with acids, that is, it has been cut into short lengths, and the fibres opened up and separated from one another.

[Footnote A: The New Explosive Co. Works.]

[Footnote B: Royal Gunpowder Factory.]

[Footnote C: Costs from £10 to £25 a ton. In his description of the "Preparation of Cotton-waste for the Manufacture of Smokeless Powder," A. Hertzog states that the German military authorities require a cotton which when thrown into water sinks in two minutes; when nitrated, does not disintegrate; when treated with ether, yields only 0.9 per cent. of fat; and containing only traces of chlorine, lime, magnesia, iron, sulphuric acid, and phosphoric acid. If the cotton is very greasy, it must be first boiled with soda-lye under pressure, washed, bleached with chlorine, washed, treated with sulphuric acid or HCl, again washed, centrifugated, and dried; if very greasy indeed a preliminary treatment with lime-water is desirable. See also "Inspection of Cotton-Waste for Use in the Manufacture of Gun-cotton," by C.E. Munro, *Jour. Am. Chem. Soc.*, 1895, 17, 783.]

~Drying the Cotton.~ — This operation is performed in either of two ways. The cotton may either be placed upon shelves in a drying house, through which a current of hot air circulates, or dried in steam-jacketed cylinders. It is very essential that the cotton should be as dry as possible before dipping in the acids, especially if a wholly "insoluble" nitro-cellulose is to be obtained. After drying it should not contain more than 0.5 per cent. of moisture, and less than this if possible. The more general method of drying the cotton is in steam-jacketed tubes, i.e., double cylinders of iron, some 5 feet long and 1-1/2 foot wide. The cotton is placed in the central chamber (Fig. 10), while steam is made to circulate in the surrounding jacket, and keeps the whole cylinder at a high temperature (steam pipes may be coiled round the outside of an iron tube, and will answer equally well). By means of a pipe which communicates with a compressed air reservoir, a current of air enters at the bottom, and finds its way up through the cotton, and helps to remove the moisture that it contains. The raw cotton generally contains about 10 per cent. of moisture and should be dried until it contains only 1/2 per cent. or less. For this it will generally have to remain in the drying cylinder for about five hours. At the end of that time a sample should be taken from the *top* of the cylinder, and dried in the water oven (100° C.[A]) for an hour to an hour and a half, and re-weighed, and the moisture then remaining in it calculated.

[Footnote A: It is dried at 180° C. at Waltham Abbey, in a specially constructed drying chamber.]

[Illustration: FIG. 10. — COTTON DRIER.]

It is very convenient to have a large copper water oven, containing a lot of small separate compartments, large enough to hold about a handful of the cotton, and each compartment numbered, and corresponding to one of the drying cylinders. The whole apparatus should be fixed against the wall of the laboratory, and may be heated by bringing a small steam pipe from the boiler-house. It is useful to have a series of copper trays, about 3 inches by 6 inches, numbered to correspond to the divisions in the steam oven, and exactly fitting them. These trays can then be taken by a boy to the drying cylinders, and a handful of the cotton from each placed in them, and afterwards brought to the laboratory and weighed (a boy can do this very well), placed in their respective divisions of the oven, and left for one to one and a half hours, and re-weighed.

When the cotton is found to be dry the bottom of the drying cylinder is removed, and the cotton pushed out from the top by means of a piece of flat wood fixed on a broom-handle. It is then packed away in galvanised- iron air-tight cases, and is ready for the next operation. At some works the cotton is dried upon shelves in a drying house through which hot air circulates, the shelves being of canvas or of brass wire netting. The hot air must pass under the shelves and through the cotton, or the process will be a very slow one.

~Dipping and Steeping.~ — The dry cotton has now to be nitrated. This is done by dipping it into a mixture of nitric and sulphuric acids. The acids used must be strong, that is, the nitric acid must be at least of a gravity of 1.53 to 1.52, and should contain as little nitric oxide as possible. The sulphuric acid must have a specific gravity of 1.84 at 15° C., and contain about 97 per cent. of the mono-hydrate (H_2SO_4). In fact, the strongest acids obtainable should be used when the product required is gun-cotton, i.e., the highest nitrate.

The sulphuric acid takes no part in the chemical reaction involved, but is necessary in order to combine with the water that is liberated in the reaction, and thus to maintain the strength of the nitric acid. The reaction which takes place is the following: —

$2(C_6H_{10}O_5) + 6HNO_3 = C_{12}H_{14}(NO_3)_6 + 6 H_2O$.

    324 378 = 594 108.
    Cellulose. Gun-Cotton.

Theoretically,[A] therefore, 1 part of cellulose should form 1.8 part of gun-cotton. Practically, however, this is never obtained, and 1.6 lb. from 1 lb. of cellulose is very good working. The mixture of acids used is generally 1 to 3, or 25 per cent. nitric acid to 75 per cent. sulphuric acid.

[Footnote A: (594 x 1)/324= 1.83.]

[Illustration: FIG. 11.—TANK FOR DIPPING COTTON.]

[Illustration: FIG. 12.—THE COOLING PITS.]

The dipping is done in cast-iron tanks (Fig. 11), a series of which is arranged in a row, and cooled by a stream of cold water flowing round them. The tanks hold about 12 gallons, and the cotton is dipped in portions of 1 lb. at a time. It is thrown into the acids, and the workman moves it about for about three minutes with an iron rabble. At the end of that time he lifts it up on to an iron grating, just above the acids, fixed at the back of the tank, where by means of a movable lever he gently squeezes it, until it contains about ten times its weight of acids (the 1 lb. weighs 10 lbs.). It is then transferred to earthenware pots to steep.

[Illustration: FIG. 13.—COTTON STEEPING POT.]

~Steeping.~—The nitrated cotton, when withdrawn from the dipping tanks, and still containing an excess of acids, is put into earthenware pots of the shape shown in Figs. 12 and 13. The lid is put on, and the pots placed in rows in large cooling pits, about a foot deep, through which a stream of water is constantly flowing. These pits form the floor of the steeping house. The cotton remains in these pots for a period of forty-eight hours, and must be kept cool. Between 18° and 19° C. is the highest temperature desirable, but the cooler the pots are kept the better. At the end of forty-eight hours the chemical reaction is complete, and the cotton is or should be wholly converted into nitro-cellulose; that is, there should be no unnitrated cotton.

[Illustration: FIG. 14.—HYDRO-EXTRACTOR.]

~Whirling Out the Acid.~—The next operation is to remove the excess of acid. This is done by placing the contents of two or three or more pots into a centrifugal hydro-extractor (Fig. 14), making 1,000 to 1,500 revolutions per minute. The hydro-extractor consists of a machine with both an inner cylinder and an outer one, both revolving in concert and driving outwardly the liquid to the chamber, from which it runs away by a discharge pipe. The wet cotton is placed around the inner cone. The cotton, when dry, is removed, and at once thrown into a large tank of water, and the waste acids are collected in a tank.[A]

[Footnote A: Care must be taken in hot weather that the gun-cotton does not fire, as it does sometimes, directly the workman goes to remove it after the machine is stopped. It occurs more often in damp weather. Dr Schüpphaus, of Brooklyn, U.S.A., proposes to treat the waste acids from the nitration of cellulose by adding to them sulphuric anhydride and nitric acid. The sulphuric anhydride added converts the water liberated from the cellulose into sulphuric acid.]

~Washing.~—The cotton has now to be carefully washed. This is done in a large wooden tank filled with water. If, however, a river or canal runs through the works, a series of wooden tanks, the sides and bottoms of which are pierced with holes, so as to allow of the free circulation of water, should be sunk into a wooden platform that overhangs the surface of the river in such a way that the tanks are immersed in the water, and of course always full. During the time that the cotton is in the water a workman turns it over constantly with a wooden paddle. A stream of water, in the form of a cascade, should be allowed to fall into these tanks. The cotton may then be thrown on to this stream of water, which, falling some height, at once carries the cotton beneath the surface of the water. This proceeding is necessary because the cotton still retains a large excess of strong acids, and when mixed with water gives rise to considerable heat, especially if mixed slowly with water. After the cotton has been well washed, it is again wrung out in a centrifugal machine, and afterwards allowed to steep in water for some time.

[Illustration: FIG. 15_a_.—THE BEATER FOR GUN-COTTON.]

~Boiling.~—The washed cotton is put into large iron boilers with plenty of water, and boiled for some time at 100° C. In some works lead-lined tanks are used, into which a steam pipe is led. The soluble impurities of unstable character, to which Sir F.A. Abel traced the liability of gun-cotton to instability, are thereby removed. These impurities consist of the products formed by the action of nitric acid on the fatty and resinous substances contained in the cotton fibres. The water in the tanks should be every now and again renewed, and after the first few boilings the water should be tested with litmus paper until they are no longer found to be acid.

[Illustration: FIG. 15_b_.—WHEEL OF BEATER.]

~Pulping.~ — The idea of pulping is also due to Abel. By its means a very much more uniform material is obtained. The process is carried out in an apparatus known as a "Beater" or "Hollander" (Fig. 15, *a, b*). It consists of a kind of wooden tank some 2 or 3 feet deep of an oblong shape, in which a wheel carrying a series of knives is made to revolve, the floor of the tank being sloped up so as to almost touch the revolving wheels. This part of the floor, known as the "craw," is a solid piece of oak, and a box of knives is fixed into it, against which the knives in the revolving wheel are pressed. The beater is divided into two parts — the working side, in which the cotton is cut and torn between the knife edges in the revolving cylinder and those in the box; and the running side, into which the cotton passes after passing under the cylinder. The wheel is generally boxed in to prevent the cotton from being thrown out during its revolution. The cotton is thus in constant motion, continually travelling round, and passing between the knives in the revolving cylinder and those in the box fixed in the wooden block beneath it. The beater is kept full of water, and the cotton is gradually reduced to a condition of pulp. The wheel revolves at the rate of 100 to 150 times a minute.

[Illustration: FIG. 16_a_. — POACHER FOR WASHING GUN-COTTON.]

[Illustration: FIG. 16_b_. — PLAN OF THE POACHER.]

[Illustration: FIG. 16_c_. — ANOTHER FORM OF POACHER.]

When the gun-cotton is judged to be sufficiently fine, the contents of the beater are run into another very similar piece of machinery, known as the "poacher" (Fig. 16, *a, b, c*), in which the gun-cotton is continuously agitated together with a large quantity of water, which can be easily run off and replaced as often as required. When the material is first run into the poacher from the beater, the water with which it is then mixed is first run away and clean water added. The paddle wheel is then set in motion, and at intervals fresh water is added. There is a strainer at the bottom of the poacher which enables the water to be drawn off without disturbing the cotton pulp. After the gun-cotton has been in the poacher for some time, a sample should be taken by holding a rather large mesh sieve in the current for a minute or so. The pulp will thus partly pass through and partly be caught upon the sieve, and an average sample will be thus obtained. The sample is squeezed out by hand, bottled, and taken to the laboratory to be tested by the heat test for purity. It first, however, requires to be dried. This is best done by placing the sample between coarse filter paper, and then putting it under a hand-screw press, where it can be subjected to a tolerably severe pressure for about three minutes. It is then rubbed up very finely with the hands, and placed upon a paper tray, about 6 inches by 4-1/2 inches, which is then placed inside a water oven upon a shelf of coarse wire gauze, the temperature of the oven being kept as near as possible to 120° F. (49° C.), the gauze shelves in the oven being kept about 3 inches apart. The sample is allowed to remain at rest for fifteen minutes in the oven, the door of which is left wide open. After the lapse of fifteen minutes the tray is removed and exposed to the air of the laboratory (away from acid fumes) for two hours, the sample being at some point within that time rubbed upon the tray with the hand, in order to reduce it to a fine and uniform state of division. Twenty grains (1.296 grm.) are used for the test. (See Heat Test, page 249.)

If the gun-cotton sample removed from the poacher stands the heat test satisfactorily, the machine is stopped, and the water drained off. The cotton is allowed some little time to drain, and is then dug out by means of wooden spades, and is then ready for pressing. The poachers hold about 2,000 lbs. of material, and as this represents the products of many hundred distinct nitrating operations, a very uniform mixture is obtained. Two per cent. of carbonate of soda is sometimes added, but it is not really necessary if the cotton has been properly washed.

~Compressing Gun-Cotton.~—The gun-cotton, in the state in which it is removed from the poacher, contains from 28 to 30 per cent. of water. In order to remove this, the cotton has to be compressed by hydraulic power. The dry compressed gun-cotton is packed in boxes containing 2,500 lbs. of dry material. In order to ascertain how much of the wet cotton must be put into the press, it is necessary to determine the percentage of water. This may be done by drying 2,000 grains upon a paper tray (previously dried at 100° C.) in the water oven at 100° C. for three hours, and re-weighing and calculating the percentage of water. It is then easy to calculate how much of the wet gun-cotton must be placed in the hopper of the press in order to obtain a block of compressed cotton of the required weight. Various forms of presses are used, and gun-cotton is sent out either as solid blocks, compressed discs, or in the form of an almost dry powder, in zinc- lined, air-tight cases. The discs are often soaked in water after compression until they have absorbed 25 per cent. of moisture.

[Illustration: FIG. 17. — OLD METHOD. 100 PIECES.]

[Illustration: FIG. 18. — NEW METHOD. ONE SOLID BLOCK.]

At the New Explosives Company's Stowmarket Works large solid blocks of gun-cotton are pressed up under a new process, whereby blocks of gun- cotton, for use in submarine mines or in torpedo warheads, are produced. Large charges of compressed gun-cotton have hitherto been built up from a number of suitably shaped charges of small dimensions (Fig. 17), as it has been impossible to compress large charges in a proper manner. The formation of large-sized blocks of gun-cotton was the invention of Mr A. Hollings. Prior to the introduction of this method, 8 or 9 lbs. had been the limit of weight for a block. This process has been perfected at the Stowmarket factory, where blocks varying from the armour-piercing shell charge of a few ounces up to blocks of compressed gun-cotton mechanically true, weighing 4 to 5 cwts. for torpedoes or submarine mines, are now produced. At the same time the new process ensures a uniform density throughout the block, and permits of any required density, from 1.4 downwards, being attained; it is also possible exactly to regulate the percentage of moisture, and to ensure its uniform distribution. The maximum percentage of moisture depends, of course, upon the density. By the methods of compression gun-cotton blocks hitherto employed, blocks of a greater thickness than 2 inches, or of a greater weight than 9 lbs., could not be made, but with the new process blocks of any shape, size, thickness, or weight that is likely to be required can be made readily and safely. The advantages which are claimed for the process may be enumerated as follows: — (1.) There is no space wasted, as in the case with built-up charges, through slightly imperfect contact between the individual blocks, and thus either a heavier charge — i.e., about 15 per cent. more gun- cotton — can be got into the same space, or less space will be occupied by a charge of a given weight. (2.) The metallic cases for solid charges may be much lighter than for those built-up, since with the former their function is merely to prevent the loss of moisture from wet gun-cotton, or to prevent the absorption of moisture by dry gun-cotton. They can thus be made lighter, as the solid charge inside will prevent deformation during transport. With built-up charges the case must be strong enough to prevent damage, either to itself or to the charge it contains. For many uses a metal case, however light, may be discarded, and one of a thin waterproof material substituted. (3.) The uniform density of charges made by this process is very favourable to the complete and effective detonation of the entire mass, and to the presence of the uniform amount of moisture in every part of the charge. (4.) Any required density, from the maximum downwards, may be obtained with ease, and any required amount of moisture left in the charge. These points are of great importance in cases where, like torpedo charges, it is essential to have the centre of gravity of the charge in a predetermined position both vertically and longitudinally, and the charge so fixed in its containing case that the centre of gravity cannot shift. The difficulty of ensuring this with a large torpedo charge built up from a number of discs and segments is well known. Even with plain cylindrical or prismatic charges a marked saving in the process of production is effected by this new system. The charges being in one block they are more easily handled for the usual periodical examination, and they do not break or chafe at the edges, as in the case of discs and cubes in built-up charges. A general view of the press is given in Fig. 19. The gun-cotton in a container is placed on a cradle fixed at an angle to the press. The mould is swivelled round, and the charge pushed into it with a rammer, and it is then swivelled back into position. The mould is made up of a number of wedge pieces which close circumferentially on the enclosed mass, which is also subjected to end pressure. Holes are provided for the escape of water.

[Illustration: FIG. 19.— A 4-CWT. BLOCK OF GUN-COTTON BEING TAKEN FROM
HYDRAULIC PRESS.]

~The Waltham Abbey Process.~ — At the Royal Gunpowder Factory, Waltham Abbey, the manufacture of gun-cotton has been carried out for many years. The process used differs but little from that used at Stowmarket. The cotton used is of a good quality, it is sorted and picked over to remove foreign matters, &c., and is then cut up by a kind of guillotine into 2-inch lengths. It is then dried in the following manner. The cotton is placed upon an endless band, which conducts it to the stove, or drying closet, a chamber heated by means of hot air and steam traps to about 180° F.; it falls upon a second endless band, placed below the first; it travels back again the whole length of the stove, and so on until delivered into a receptacle at the bottom of the farther end, where it is kept dry until required for use. The speed at which the cotton travels is 6 feet per minute, and as the length of the band travelled amounts to 126 feet, the operation of drying takes twenty-one minutes. One and a quarter lb. are weighed out and placed in a tin box; a truck, fitted to receive a number of these boxes, carries it along a tramway to a cool room, where it is allowed to cool.

~Dipping.~ — Mixed acids are used in the proportion of 1 to 3, specific gravity nitric acid 1.52, and sulphuric acid 1.84. The dipping tank is made of cast iron, and holds 220 lbs. of mixed acids, and is surrounded on three sides by a water space in order to keep it cool. The mixed acids are stored in iron tanks behind the dipping tanks, and are allowed to cool before use. During the nitration, the temperature of the mixed acids is kept at 70° F., and the cotton is dipped in quantities of 1-1/2 lb. at a time. It is put into a tin shoot at the back of the dipping tank, and raked into the acids by means of a rabble. It remains in the acids for five or six minutes, and is then removed to a grating at the back, pressed and removed. After each charge of cotton is removed from the tank, about 14 lbs. of fresh mixed acids are added, to replace amount removed by charge. The charge now weighs, with the acids retained by it, 15 lbs.; it is now placed in the pots, and left to steep for at least twenty-four hours, the temperature being kept as low as possible, to prevent the formation of soluble cotton, and also prevent firing. The proportion of soluble formed is likely to be higher in hot weather than cold. The pots must be covered to prevent the absorption of moisture from the air, or the accidental entrance of water, which would cause decomposition, and consequent fuming off, through the heat generated by the action of the water upon the strong acids.

The excess of acids is now extracted by means of hydro-extractors, as at Stowmarket. They are worked at 1,200 revolutions per minute, and whirled for five minutes (10-1/2 lbs. of waste acids are removed from each charge dipped). The charge is then washed in a very similar manner to that previously described, and again wrung out in a centrifugal extractor (1,200 revolutions per minute). The gun-cotton is now boiled by means of steam in wooden tanks for eight hours; it is then again wrung out in the extractors for three minutes, boiled for eight hours more, and again wrung out; it is then sent to the beater and afterwards to the poacher. The poachers hold 1,500 gals. each, or 18 cwt. of cotton. The cotton remains six hours in the poachers. Before moulding, 500 gals. of water are run into the poacher, and 500 gals. of lime water containing 9 lbs. of whiting and 9 gals. of a caustic soda solution. This mixture is of such a strength that it is calculated to leave in the finished gun-cotton from 1 to 2 per cent. of alkaline matter.

By means of vacuum pressure, the pulp is now drawn off and up into the stuff chest—a large cylindrical iron tank, sufficiently elevated on iron standards to allow room for the small gauge tanks and moulding apparatus below. It holds the contents of one poacher (18 cwt.), and is provided with revolving arms to keep the pulp stirred up, so that it may be uniformly suspended in water.

Recently a new process, invented by J.M. and W.T. Thomson (Eng. Pat. No. 8,278, 1903), has been introduced at the Waltham Abbey Factory. The object of this invention is the removal of the acids of nitration from the nitrated material after the action has been completed, and without the aid of moving machinery, such as presses, rollers, centrifugals, and the like. The invention consists in the manufacture of nitrated celluloses by removing the acids from the nitrated cellulose directly by displacement without the employment of either pressure or vacuum or mechanical appliances of any kind, and at the same time securing the minimum dilution of the acids. It was found that if water was carefully run on to the surface of the acids in which the nitro-cellulose is immersed, and the acids be slowly drawn off at the bottom of the vessel, the water displaces the acid from the interstices of the nitro-cellulose without any undesirable rise in temperature, and with very little dilution of the acids. By this process almost the whole of the acid is recovered in a condition suitable for concentration, and the amount of water required for preliminary washing is very greatly reduced. The apparatus which is used for the purpose consists of a cylindrical or rectangular vessel constructed with a perforated false bottom and a cock at its lowest point for running off the liquid. Means are also provided to enable the displacing water to be run quietly on to the surface of the nitrating acids.[A]

[Footnote A: In a further patent (Eng. Pat. 7,269, 1903, F.L. Natham), J.M. Thomson and W.T. Thomson propose by use of alcohol to replace the water, used in washing nitro-cellulose, and afterward to remove the alcohol by pressing and centrifuging.]

The apparatus is shown in Fig. 20, side elevation, and in Fig. 21 a plan of the nitrating vessel and its accessories is given. In Fig. 20 is shown in sectional elevation one of the trough devices for enabling liquids to be added to those in the nitrating vessel without substantial disturbance.

[Illustration: FIG. 20.—SECTIONAL ELEVATION OF THOMSON'S APPARATUS, a, Tank; b, False Bottom; c, Bottom; c', Ribs; d, Draining Outlet; e, Grid; f, Troughs, with Aprons g; h, Pipe, with Branches h', leading to Troughs, f; k', Outlet Pipe of the Sulphuric Acid Tank k; l, Water Supply Pipe; m, Pipe to supply of Nitrating Acids; o, Perforations of Trough f; p, Cock to remove Acid.]

In carrying out this invention a rectangular lead-lined or earthenware tank a is employed, having a false bottom b, supported by ribs c', over the real bottom c, which slopes down to a draining outlet pipe d, provided with a perforated grid or plate e, adapted to prevent choking of the outlet. Suitably supported near the top of the vessel a are provided two troughs, f having depending aprons g, a pipe h has two branches h', leading to the troughs, f. This pipe h is adapted to be connected by a rubber pipe either to the outlet pipe k' of the sulphuric acid tank k or the water supply pipe l. The nitrating acids are supplied through the pipe m. A charge of mixed nitrating acids is introduced into the vessel a say up to the level n, and the dry cellulose thrown into the acids in small quantities at a time, being pushed under the surface in the usual way.

[Illustration: FIG. 21. — PLAN OF THOMSON'S APPARATUS, a, Tank; b,
False Bottom; c', Ribs; e, Grid; f, Troughs; g, Aprons; h and h', Pipes to Troughs f; k, Sulphuric Acid Tank; m, Pipe to Nitrating Acids Tank; o, Perforations of Troughs; p, Cock to remove Acid.]

A thin layer, say half an inch, of a suitable liquid, preferably sulphuric acid, of a gravity not exceeding that of the waste acid to be produced, is run carefully on the top of the acids by means of the troughs f, which are perforated as shown at o, so that the sulphuric acid runs down the aprons g, and floats on the nitrating acids. The whole is then allowed to stand till nitration has been completed. Water is then supplied to the troughs by way of the pipes l, h, and h', and is allowed to float very gently over the surface of the sulphuric acid, and when a sufficient layer has been formed, the cock p at the bottom of the apparatus is opened, and the acid slowly drawn off, water being supplied to maintain the level constant. It is found that the rate of displacement of the acids is a factor which exerts a considerable influence on the properties of the resulting nitro-cellulose, and affords a means of regulating the temperature of displacement. A rate of displacement which has been found suitable is about two inches in depth of the vessel per hour when treating highly nitrated celluloses, but this rate may, in some cases, be considerably increased. The flow of water at the top of the apparatus is regulated so that a constant level is maintained. By this means the water gradually and entirely displaces the acids from the interstices of the nitro-cellulose, the line of separation between the acids and the water being fairly sharply defined throughout. The flow of water is continued until that issuing at the bottom is found to be free from all trace of acid. The purification of the nitro-cellulose is then proceeded with as usual, either in the same vessel or another.

In the process above described, the object of the introduction of a small layer of sulphuric acid is mainly to prevent the fuming which would otherwise take place, and is not essential, as it is found it can be omitted without any deleterious effect. In order to use the mixed acids in the most economical manner, the waste acid from a previous operation may be used for a first nitration of the cellulose; being afterwards displaced with fresh acids which carry the nitration to the required degree before they are in turn displaced by water. The apparatus may be used merely for the removal of the acid, in which case the nitration is carried out in other vessels in the usual way, and the nitro-cellulose removed to the displacement apparatus where it is just covered with waste acid, and the displacement then proceeded with as above described. In some cases the process is carried out in an ordinary nitrating centrifugal, using the latter to effect preliminary drying after acid extraction. This gives a great advantage over the usual method of working ordinary centrifugal nitrating apparatus, because the acid being removed before the centrifugal is run, practically all danger of firing therein disappears, and a greater proportion of the waste acid is recovered.

In some cases the acids and water may be supplied by perforated pipes, lying along the edges of the nitrating vessel, and these edges may, if desired, be themselves made inclined, like the sides of the troughs f. In the case of effecting nitration in centrifugals as above, the displacing sulphuric acid and water may thus be supplied round the edges of the machines, or removal troughs such as f may be used. It will be obvious that any inert liquid of suitable specific gravity may be used instead of sulphuric acid, as a separation layer.

~Moulding.~ — By means of the small measuring tank above referred to, the gun-cotton pulp is drawn off from the stuff chest, and run into moulds of the shapes and sizes required. Thence a large proportion of the water is drawn off by means of tubes connected with the vacuum engine, the moulds having bottoms of fine wire gauze, in order to prevent the pulp from passing through. Hydraulic pressure of about 34 lbs. on the square inch is then applied, which has the effect of compressing the pulp into a state in which it has sufficient consistency to enable it to be handled with care, and also expels a portion of the remaining water.

~Compressing.~ — The moulded gun-cotton is now taken to the press house, which is situated at some distance from the rest of the factory. Here the moulds are subjected to powerful hydraulic pressure, from 5 to 6 tons per square inch, and is compressed to one-third of its previous bulk. The slabs or discs thus formed are kept under pressure for a short time, not exceeding a minute and a half, to give the requisite density. It should, when removed, be compact, and just sink in water, and should perceptibly yield to the pressure of the fingers. There are perforations in the press blocks, to allow of the escape of gases, if formed, by reason of sufficient heat being generated. The men working the press are placed under cover, behind strong rope mantlets having eye tubes which command a view of the press.

~Packing.~ — The finished slabs and discs are dipped into a solution of soda and carbolic acid, and packed in special wood metal-lined cases. When it is to be sent abroad, the metal lining, which is made of tinned copper, is soldered down, but both the outer wooden and inner metal cases are fitted with air-tight screw-plugs, so that when necessary water can be added without unfastening the cases.

~Reworked gun-cotton~ does not make such good discs as new pulped gun- cotton, probably because the fibrous tenacity of the gun-cotton has been destroyed by the amount of pressure it has previously undergone, so that when repulped it resembles fine dust, and a long time is required to press it into any prescribed form. It is generally boiled for eight hours to open up the fibre and remove alkali, then broken up by hand with wooden mallets, pulped, and then used with fresh gun-cotton in the proportion of 1 to 5 parts.

~Manufacture at Le Bouchet.~ — At Le Bouchet gun-cotton was made thus: — 200 grms. of cotton were steeped for an hour in 2 litres of a mixture of 1 volume concentrated nitric and 2 volumes sulphuric acid. The cotton was then removed and pressed, whereby 7/10ths of the waste acids was recovered. After this it was washed for one to one and a half hours in running water, strongly pressed again; allowed to lie for twenty-four hours in wood-ash lye; then well washed in running water; pressed, and finally dried on a wide linen sheet, through which was forced air heated to 60° C. The average yield from 100 parts of cotton was 165 parts of gun- cotton. The strong pressings of the gun-cotton, while still impregnated with acids, caused subsequent washings to be difficult and laborious.

~Granulation of Gun-Cotton.~—Gun-cotton is often required in the granulated form for use either alone or with some form of smokeless powder. This is done under the patent of Sir Frederick Abel in the following manner:—The gun-cotton from the poacher is placed in a centrifugal machine, very similar to the hydro-extractors before mentioned, and used for wringing out the acids. In this machine it loses water until it only contains 33 per cent., and is at the same time reduced to a more or less fibrous state. It is then taken to the granulating room, where it is first passed through sieves or perforations, which break up the mass into little pieces like shot. The material is then transferred to a revolving drum made of wood or stout leather, which is kept constantly revolving for some time. The material is occasionally sprinkled with water. The drum in turning, of course, carries the granules partially round with it, but the action of gravity causes them to descend constantly to the lowest point, and thus to roll over one another continually. The speed of the drum must not be too rapid. None of the granules must be carried round by centrifugal force, but it must be fast enough to carry them some little distance up the side of the drum. After removal from the drum the granules are dried upon shelves in the drying house.

Gun-cotton is also dissolved in acetone or acetic ether until it has taken the form of a jelly. It is then rolled into thin sheets, and when dry cut up into little squares. In the manufacture of smokeless powders from nitro-cellulose, nitro-lignine, &c., the various substances are mixed with the gun-cotton or collodion-cotton before granulating.

~Collodion-Cotton.~—In the manufacture of collodion or soluble cotton the finer qualities of cotton-waste are used and the acids used in the dipping tanks are much weaker. The manufacture of collodion-cotton has become of more importance than gun-cotton, by reason of its use for the manufacture of the various forms of gelatine, such as gelatine dynamite, gelignite, forcite, &c., and also on account of its extensive use in the manufacture of many of the smokeless powders. It is also used for the manufacture of "collodion," which is a solution of collodion-cotton in ether-alcohol; for the preparation of celluloid, and many other purposes. It is less explosive than gun-cotton, and consists of the lower nitrates of cellulose. It is soluble in nitro-glycerine, and in a mixture of 2 parts of ether and 1 of alcohol; also in acetone, acetic ether, and other solvents. MM. Ménard and Domonte were the first to prepare a soluble gun-cotton, and its investigation was carried on by Béchamp, who showed that its properties and composition were different to those of gun-cotton.

~Manufacture.~—The cotton used is cotton-waste.[A] It is thought by some that Egyptian cotton is preferable, and especially long fibre varieties. The strength of the acids used is, however, of more importance than the quality of the cotton. The percentage composition of the acid mixture which gives the best results is as follows:—Nitric acid, 23 per cent.; sulphuric acid, 66 per cent.; and water, 11 per cent; and has a specific gravity of 1.712 (about). It can be made by mixing sulphuric acid of specific gravity 1.84 with nitric acid of specific gravity 1.368 in the proportions of 66 per cent. and 34 per cent. respectively. (The production of the penta-nitro-cellulose is aimed at if the collodion-cotton is for use as an explosive.) If the acids are much weaker than this, or potassium nitrate and sulphuric acid is used, the lower nitrates will be formed. The product, while being entirely soluble in ether-alcohol or nitro-glycerine, will have a low nitrogen content, whereas a material with as high a nitrogen as 12 or 12.6 is to be aimed at.

[Footnote A: Raw cotton is often used.]

The cotton should not be allowed to remain in the dipping tanks for more than five minutes, and the acid mixture should be kept at a temperature of 28° C. or thereabouts; and the cotton should be removed after a few minutes, and should not be pressed out, as in the case of gun-cotton, but at once transferred to the pots and allowed to steep for forty-eight hours. (Some prefer twenty-four hours, but there is more chance in this case of the product containing non-nitrated cellulose.) When the nitration is complete, the collodion-cotton is removed from the pots, and treated in exactly the same manner as described under gun-cotton. The produce should be entirely soluble in ether-alcohol and nitro-glycerine, and contain as near 12.7 per cent. of nitrogen as possible. The theoretical nitrogen is for the penta-nitro-cellulose 12.75 per cent. This will, however, seldom if ever be obtained. The following are some of the results I have obtained from different samples:—

        Nitrogen.
   (1.) (2.) (3.)
German make 11.64 11.48 11.49 per cent.
Stowmarket 12.57 12.60 11.22 "
Walsrode 11.61 12.07 11.99 "
Faversham 12.14 11.70 11.60 "

and the following was the analysis of a sample (No. 1) of German-made collodion-cotton, which made very good blasting gelatine:—

Soluble cotton (collodion) 99.118 per cent. | *Nitrogen = 11.64 per cent.*
Gun-cotton 0.642 " |
Non-nitrated cotton 0.240 "
Total ash 0.25 "

It should contain as little non-nitrated or unconverted cotton and as little gun-cotton as possible, as they are both insoluble in nitroglycerol. The quality and composition of any sample of collodion-cotton can be quickly inferred by determining the percentage of nitrogen by means of the nitrometer and the use of the solubility test.[A] A high nitrogen content coupled with a high solubility is the end to be aimed at; a high nitrogen with a low solubility shows the presence of gun-cotton, and a low nitrogen, together with a low solubility, the presence of unnitrated cotton. Where complete solubility is essential and the percentage of nitrogen less important, Dr Lunge recommends nitration with a mixture of equal parts of sulphuric and nitric acids containing from 19 to 20 per cent. of water.

[Footnote A: See Analysis of Explosives.]

Mr T.R. France claims to have invented some improvements in the manufacture of soluble nitro-cellulose. His object has been to produce an article as uniform as possible. His explanation of the imperfect action of the acids is that, however uniform the mixed acids may be in strength and proportions, and however carefully the operations of nitrating, &c., may be conducted, there are variable elements found in different samples of cotton. The cotton fibre has for its protection a glazed surface. It is tubular and cellular in structure, and contains a natural semi-fluid substance composed of oil or gum, which varies in nature according to the nature of the soil upon which the cotton is grown. The tubes of the fibre seem to be open at one end only when the fibre is of normal length. When, therefore, the cotton is subjected to the action of the mixed acids, the line of least resistance seems to be taken by them, viz., the insides of the tubes constituting the fibre of the cotton, into which they are taken by capillary attraction, and are subject to change as they progress, and to the increased resistance from the oil or gum, &c., in their progress, and therefore to modified action, the result of which is slower and slower action, or chemical change. He also thinks it is possible that the power of capillary attraction is balanced in the tubes by air contained therein, after a little, sufficiently so to prevent the acids from taking full effect. To get over this, Mr France uses his cotton in a fine state, almost dust, in fact, and then nitrates in the usual mixture of acids at 40° to 90° F., the excess of acids being removed by pressure. He says he does not find it necessary to wash this fine cotton dust in an alkaline solution previous to nitration. His mixed acids consist of 8 parts HNO_3 = 42° B., and 12 parts H_2SO_4 = 66° B., and he stirs in the dipping tank for fifteen minutes, the temperature being 50° F. to 100° F., the temperature preferred being 75° F.

~"Nitrated" Gun-Cotton.~ — The nitrates that are or have been mixed with gun-cotton in order to supply oxygen are potassium nitrate, ammonium nitrate, and barium nitrate (tonite). The total combustion of gun-cotton by potassium nitrate corresponds to the equation: —

$$10[C_{24}H_{18}(NO_3H)_{11}O_9] + 82KNO_3 = 199CO_2 + 41K_2CO_3 + 145H_2O + 96N_2,$$

or 828 grms. of nitrate for 1,143 grms. of gun-cotton, or 42 per cent. nitrate and 58 per cent. gun-cotton. The explosive made at Faversham by the Cotton Powder Company, and known as tonite No. 1, consists of very nearly half gun-cotton and half barium nitrate. The relations by weight of total combustion would be 51.6 of gun-cotton to 48.4 of barium nitrate. The average composition of tonite I have found by analysis to be 51 per cent. gun-cotton to 49 per cent. barium nitrate. The heat liberated is practically the same as for an equivalent weight of KNO_3; but the barium nitrate mixture weighs 2,223 grms. instead of 1,971 grms., or one-eighth more. The advantage in mixing a nitrate with gun-cotton is that it supplies oxygen, and by converting all the carbon into carbonic acid, prevents the formation of the poisonous gas carbonic oxide (CO). The nitrates of potassium and barium are also used admixed with nitrocellulose in several of the sporting smokeless powders.

~The Manufacture of Tonite.~ — The explosive tonite was patented by Messrs Trench, Faure, and Mackie, and is manufactured at Faversham and Melling at the works of the Cotton Powder Company, and at San Francisco by the Tonite Powder Company. It consists of finely divided and macerated gun-cotton incorporated with finely ground nitrate of barium which has been carefully recrystallised. It is made by acting upon carbonate of barium[A] with nitric acid. The wet and perfectly purified, finely pulped gun-cotton is intimately mixed up between edge runners with about the same weight of nitrate, and the mixing and grinding continued until the whole has become an intimately mixed paste. This paste is then compressed into cartridges, formed with a recess at one end for the purpose of inserting the detonator. The whole is then covered with paraffined paper.

[Footnote A: Witherite, $BaCO_3 + 2HNO_3 = Ba(NO_3)_2 + CO_2 + H_2O$.]

The tonite No. 2 consisted of gun-cotton, nitrates of potash and soda, charcoal and sulphur. Tonite No. 3[A] is composed as follows: — Gun-cotton, 19 per cent.; di-nitro-benzol, 13 per cent.; and barium nitrate, 68 per cent. or similar proportions. It is a yellowish colour, and being slower in its explosive action, is better adapted for blasting soft rock.

[Footnote A: Tonite No. 1 was patented by Messrs Trench, Faure, and

Mackie, and tonite Nos. 2 and 3 by Trench alone.]

Tonite is extensively used in torpedoes and for submarine blasting, also for quarries, &c. Large quantities were used in the construction of the Manchester Ship Canal. Among its advantages are, that the English railways will take tonite on the same footing as gunpowder; it is a very dense material; if wetted it can easily be dried in the sun; it very readily explodes by the use of a proper detonator; while it burns very slowly and without the least danger; the cartridges being waterproofed, it can be employed in wet bore holes, and it can be tamped with water; and finally, as it contains sufficient oxygen to oxidise the carbon, no carbonic oxide (CO) gas is formed, i.e., its detonation is perfect. It is a very safe explosive to use, being little susceptible to either blows or friction.

Not long ago, a committee, composed of Prof. P. Bedson, Drs Drummond and Hume, Mr T. Bell, one of H.M. Inspectors of Coal Mines, and others, in considering the problem whether the fumes produced by the combustion of tonite were injurious to health, carried out a series of experiments in coal mines for this purpose. The air at the "intake" was analysed, also the air of the "return," and the smoky air in the vicinity of the shot holes. The cartridge was surrounded by the flame-extinguishing mixture, and packed in a brown paper bag. During the first experiment nineteen shots were fired (= 6.29 lbs. tonite). The "return" air showed only a trace of carbonic oxide gas (CO). At the second experiment thirteen shots were fired (= 4.40 lbs. tonite), and analysis of the air of the "return" showed that CO was present in traces only, whilst the fumes contained only 1.9 to 4.8 parts per 10,000.

~Dangers in connection with the Manufacture of Guncotton, &c.~—Of all the nitro compounds, the least dangerous to manufacture are gun-cotton and collodion-cotton. The fact that the Stowmarket Factory is within five minutes' walk of the town shows how safe the manufacture of this explosive is regarded. With the exception of the nitration and the compression into blocks or discs, the whole process is worked with a large excess of water, and the probability of an explosion is thus reduced to a minimum. Among the precautions that should, however, be taken, are—first, the careful extraction of the resinous and soluble substances from the cotton before nitration, as it was shown many years ago by Sir F.A. Abel that the instability of the gun-cotton first manufactured in England and Austria was chiefly due to these compounds. They are generally removed by boiling the cotton in a soda solution.

The actual nitration of cotton is not a dangerous operation, but the operations of wringing in the hydro-extractors, and washing the nitro- cotton after it leaves the first centrifugal machine, are somewhat so. Great care should be taken that the wrung-out nitro-cotton at once comes in contact with a large excess of water, i.e., is at once immersed entirely in the water, since at this stage it is especially liable to decomposition, which, once started, is very difficult to stop. The warmer the mixture and the less water it contains, the more liable it is to decomposition; hence it is that on warm and damp days the centrifugal machines are most likely to fire. The commencement of decomposition may be at once detected by the evolution of red fumes. Directly the gun-cotton is immersed in the large quantity of water in the beater and poacher it is safe.

In order that the final product may be stable and have good keeping qualities, it is necessary that it should be washed completely free from acid. The treatment in the beater and poacher, by causing the material to assume the state of a fine pulp, in contact with a large quantity of water, does a good deal to get rid of the free acid, but the boiling process is absolutely necessary. It has been proposed to neutralise the free acid with a dilute solution of ammonia; and Dr C.O. Weber has published some experiments bearing upon this treatment. He found that after treatment with ammonia, pyroxyline assumed a slightly yellowish tinge, which was a sure sign of alkalinity. It was then removed from the water, and roughly dried between folds of filter paper, and afterwards dried in an oven at 70° C. After three hours, however, an explosion took place, which entirely destroyed the strong copper oven in which the nitro- cotton (about one oz.) had been drying. The explosion was in some respects remarkable. The pyroxyline was the di-nitro-cellulose (or possibly the penta-nitro?), and the temperature was below the igniting point of this material (40° C. would have been a better temperature). Dr Weber determined the ignition point of his di-nitro-cellulose, and found it to be 194° to 198° C., and he is therefore of opinion that the explosion was due to the treatment of the partially washed material with ammonia. A certain quantity of ammonium nitrate was probably formed, and subsequently dried upon the nitro-cellulose, in a state of very fine subdivision. The faintest trace of acid would then be sufficient to bring about the explosive ignition of the ammonium nitrate.

The drying of gun-cotton or collodion-cotton is also a somewhat dangerous operation. A temperature of 40° C. (104° F.) should not be exceeded, and thermometers should be placed in the nitro-cotton, and the temperature frequently observed. An electric alarm thermometer is also a useful adjunct to the cotton drying house. Great care must also be taken that there are no exposed hot-water pipes or stoves in the drying house, as the fine gun-cotton dust produced by the turning or moving of the material upon the shelves would settle upon such pipes or stoves, and becoming hot, would be very sensitive to the least friction. The floor also should be covered with linoleum or indiarubber. When hot currents of air are made to pass over the surface of gun-cotton, the gun-cotton becomes electrified. It is important, therefore, to provide some means to carry it away. Mr W.F. Reid, F.I.C., was the first to use metal frames, carriers, and sieves, upon which is secured the cloth holding the gun-cotton, and to earth them.

The compression of gun-cotton into blocks, discs, &c., is also attended with considerable risk. Mr O. Guttmann, in an interesting paper upon "The Dangers in the Manufacture of Explosives" (*Jour. Soc. Chem. Ind.*, No. 3, vol. xi., 1892), says: "The compression of gun-cotton into cartridges requires far more care than that of gunpowder, as this is done in a warm state, and gun-cotton even when cold, is more sensitive than gunpowder. When coming out of the centrifugal machines, the gun-cotton should always pass first through a sieve, in order to detect nails or matches which may by chance have got into it. What has been said as to gunpowder presses applies still more to those for gun-cotton, although the latter are always hydraulic presses. Generally the pistons fit the mould perfectly, that is to say, they make aspiration like the piston of a pump. But there is no metal as yet known which for any length of time will stand the constant friction of compression, and after some time the mould will be wider in that part where the greatest compression takes place. The best metal for this purpose has proved to be a special steel made by Krupp, but this also is only relatively better; for pistons I prefer hard cast iron. If the position of the moulds and pistons is not exactly the same in all cases, what the Germans call 'Ecken' (English 'binding') will take place, viz., the mould will stand obliquely to the piston, and a dangerous friction will result." "Of course, it is necessary to protect the man working the hydraulic valves during compression. At Waltham Abbey they have a curtain made of ship's hawsers, which is at the same time elastic and resistant." Mr Guttmann has found that a partition wall 12 inches thick, made of 2-inch planks, and filled with ground cinders, gives very effective protection. A door in this partition enables the workman to get to the press, and a conical tube penetrates the wall, enabling the man to see the whole work from a safe standpoint. The roof, or one side of the building, should be of glass, so as to give the explosion a direction.

~Trench's Fire-extinguishing Compound~ is manufactured by the Cotton Powder Company at Faversham, and is the invention of Mr George Trench, F.C.S., the manager of the Company. The object of the invention is to surround the cartridges of tonite, when used in coal mines, with a fire- extinguishing compound. If a charge of tonite, dynamite, or gelatine dynamite is put inside a few ounces of this mixture, and then fired, not the least trace of flame can be observed, and experiments appear to show that there is no flame at all. The compound consists of sawdust impregnated with a mixture of alum and chlorides of sodium and ammonia. Fig. 22 shows the manner of placing the tonite cartridge in the paper bag, and surrounding it with the fire-extinguishing compound, *aa*. The attachment of the fuse and detonator is also shown.

[Illustration: FIG. 22. — TRENCH'S FIRE-EXTINGUISHING CARTRIDGE.]

The following report (taken from the *Faversham News*, 22nd Oct. 1887) of experiments conducted in the presence of several scientific and mining men will show its value: — "A large wrought-iron tank, of 45 cubic feet capacity, had been sunk level with the ground in the middle of the yard; to this tank the gas had been laid on, for a purpose that will be explained later on. The charges were fired by means of electricity, a small dynamo firing machine being placed from 30 to 40 yards away from the 'mine.'" Operations were commenced by the top of the tank being covered over and plastered down in order to make it air-tight; then a sufficient quantity of coal gas was placed in it to make it highly inflammable and explosive, the quantity being ascertained by a meter which had been fixed specially for the purpose. Whilst the gas was being injected the cartridge was prepared.

The first experiment was to try whether a small charge of tonite — fired without the patent extinguisher — would ignite the gas. The gas having been turned on, a miner's lamp was placed in the "tank," but this was extinguished before the full quantity of gas had gone through the meter. However, the gas being in, the charge of 1-1/4 oz. tonite was placed in the "mine," the detonator was connected by means of long wires to the dynamo machine, and the word was given to "fire." With a tremendous report, and a flash of fire, the covering of the mine flew in all directions, clearly showing that the gas had exploded. The next cartridge (a similar charge) was prepared with the patent compound. First of all a brown paper case of about 2 inches diameter was taken, and one of the tonite cartridges was placed in the centre of it, the intervening space between the charge and the case being packed with the "fire-extinguishing compound." The mine having had another supply of gas injected, the protected cartridge was placed inside and fired. The result was astonishing, the explosion not being nearly so loud, whilst there was not the least flash of fire. "Protected" and "unprotected" charges were fired at intervals, gas being turned into the tank on each occasion. Charges of tonite varying from 1 to 6 oz. were also used with the compound. The report was trifling, whilst no flash could be seen.

~Uses of Collodion-Cotton.~ — The collodion or soluble gun-cotton is used for a variety of purposes. The chief use is, however, for the manufacture of the various explosive gelatine compounds, of which blasting gelatine is the type. It is also very extensively used in the manufacture of smokeless powders, both military and sporting — in fact, very few of them do not contain it. In some, however, nitro-lignose or nitrated wood is used instead. This, however, is chemically the same thing, viz., nitro- cellulose, the cellulose being derived from the wood fibre. It is more used in this connection than the higher nitrate gun-cotton. Another use to which it has been applied very extensively, of recent years, is in the manufacture of "celluloid." It is used in photography for the preparation of the films on the sensitised plates, and many other purposes. Dissolved in a solution of two parts ether and one of alcohol, it forms the solution known as collodion, used for a variety of purposes, such as a varnish, as a paint for signals; in surgery, for uniting the edges of wounds.

Quite lately, Mr Alfred Nobel, the well-known inventor of dynamite, has patented the use of nitro-cellulose, hydro- or oxy-cellulose, as an artificial substitute for indiarubber. For this purpose it is dissolved in a suitable non-volatile or slightly volatile "solvent," such as nitro- naphthalene, di-nitro-benzene, nitro-toluene, or its homologues; products are obtained varying from a gelatinous consistency to the hardness of ebonite. The proportions will vary from about 20 per cent. of nitro- cellulose in the finished product, forming a soft rubber, to 50 per cent. nitrating celluloid, and the "solvent" chosen will depend on the use to which the rubber substitute is to be put, the liquids giving a more elastic substance, whilst mixtures of solids and liquids may be employed when the product is to be used at high temperatures. By means of rollers steam heated, the incorporation may be accomplished without the aid of a volatile liquid, or the nitro-cellulose may be employed wet, the water being removed after "solution."

It is advisable to use the cellulose nitrated only just enough to render it suitable, in order to reduce the inflammability of the finished product. Mr W. Allen, M.P., of Gateshead, proposed to use celluloid for cartridge cases, and thus to lighten ammunition, and prevent jambing, for the case will be resolved into gases along with the powder. Extractors will also be done away with.

~Celluloid~ is an intimate mechanical mixture of pyroxyline (gun-cotton or collodion-cotton) with camphor, first made by Hyatt, of Newark, U.S.A., and obtained by adding the pyroxyline to melted camphor, or by strongly compressing the two substances together, or by dissolving the constituents in an appropriate solvent, e.g., alcohol or ether, and evaporating to dryness. A combination of the two latter methods, i.e., partial solution, with pressure, is now usually adapted. The pyroxyline employed is generally the tetra- and penta-nitrated cellulose, the hexa-nitrate (gun-cotton) being but seldom used on account of its explosive properties.

Care is taken to prevent the formation of the hexa-nitrate by immersing the cellulose in only moderately strong nitric acid, or in a warm mixture of nitric and sulphuric acids. The paper, either in small pieces or in sheets, is immersed for about twenty-five minutes in a mixture of 2 parts of nitric acid and 5 parts of sulphuric acid, at a temperature of about 30° C., after which the nitrated cellulose is thoroughly washed with water to remove the last traces of free acid, pressed, and whilst still moist, mixed with the camphor.

In the process of Trebouillet and De Besancele, the cellulose, which may be in the form of paper, cotton, or linen, is twice nitrated—first in the acid mixture employed in a previous operation; and secondly, in a fresh mixture of 3 parts sulphuric acid of 1.83 specific gravity, and 2 parts concentrated nitric acid containing nitrous acid. After each nitration the mass is subjected to pressure, and is then carefully washed with water, to which, at the last, a small quantity of ammonia or caustic soda is added to remove the final traces of acid. The impregnation of the pyroxyline with the camphor is effected in a variety of ways.

The usual proportion of the constituents is 2 parts pyroxyline and 1 part camphor. In Trebouillet and De Besancele's process, 100 parts of pyroxyline are intimately mixed with from 40 to 50 parts camphor, and moulded together by strong pressure in a hot press, and afterwards dried by exposure to air, desiccated by calcium chloride or sulphuric acid. The usual method is, however, to dissolve the camphor in the least possible quantity of alcohol, and sprinkle the solution over the dry pyroxyline, which is then covered with a second layer of pyroxyline, and the whole again treated with the camphor solution, the addition of pyroxyline and camphor solution being repeated alternately until the requisite amount of celluloid mixture is obtained.

The mass, which sinks together in transparent lumps, is worked for about an hour between cold iron rollers, and then for the same period between rollers which can be gently heated by steam. The layer of celluloid surrounding the rollers is then cut away and again pressed, the resulting cake, which is now about 1 cm. thick, being cut into plates of about 70 cm. long and 30 cm. broad. These are placed one above the other, and strongly pressed together by hydraulic pressure at a temperature of about 70° for twenty-four hours. The thick cakes are once more cut into plates of the desired thickness, and placed in a chamber heated from 30° to 40° for eight to fourteen days, whereby they become thoroughly dry, and are readily made into various articles either by being moulded while warm under pressure, cut, or turned. Occasionally other liquids, e.g., ether and wood spirit, are used in place of alcohol as solvents for the camphor.

Celluloid readily colours, and can be marbled for manufacturing purposes, &c. It is highly inflammable and not explosive even under pressure, and may be worked under the hammer or between rollers without risk. It softens in boiling water, and may be moulded or pressed. Its specific gravity varies slightly with its composition and with the degree of pressure it has received. It is usually 1.35. It appears to be merely a mixture of its components, since by treatment with appropriate solvents the camphor may be readily extracted, and on heating the pyroxyline burns away while the camphor volatilises.

The manufacture of pyroxyline for the purpose of making celluloid has very much increased during recent years, and with this increase of production improved methods of manufacture have been invented. A series of interesting papers upon the manufacture of pyroxyline has been published by Mr Walter D. Field, of New York, in the *Journal of the American Chemical Society*[A] from which the following particulars are taken: —

[Footnote A: Vol. xv., No. 3, 1893; Vol. xvi., No. 7, 1894; Vol. xvi., No. 8, 1894. Figs. 19, 20, 21, 22, and 23 are taken from Mr Field's paper.]

~Selection of the Fibre.~ — Cotton fibre, wood fibre, and flax fibre in the form of raw cotton, scoured cotton, paper, and rags are most generally used, and give the best results. As the fibres differ greatly in their structure, they require different methods of nitrating. The cotton fibre is a flattened hollow ribbon or collapsed cylindrical tube, twisted a number of times, and closed at one end to form a point. The central canal is large, and runs nearly to the apex of the fibre. Its side walls are membraneous, and are readily penetrated by the mixed acids, and consequently the highest nitration results. In the flax fibre the walls are comparatively thick, the central canal small; hence it is to be presumed that the nitration must proceed more slowly than in the case of cotton. The New Zealand flax gives the most perfectly soluble nitrates of any of the flaxes. Cotton gives a glutinous collodion, and calico a fluid collodion. One of the largest manufacturers of pyroxyline in the States uses the "Memphis Star" brand of cotton. This is an upland cotton, and its fibres are very soft, moist, and elastic. Its colour is light creamy white, and is retained after nitration. The staple is short, and the twist inferior to other grades, the straight ribbon-like filaments being quite numerous. This cotton is used carded, but not scoured. This brand of cotton contains a large quantity of half and three-quarter ripe fibre, which is extremely thin and transparent, distributed throughout the bulk of the cotton (Monie., Cotton Fibre, 67). Mr Field says, "This is a significant fact when it is known that from this cotton an extremely soluble pyroxyline can be produced."

Pyroxyline of an inferior grade as regards colour only can be produced from the cotton wastes of the trade. They must be scoured before they are fit for nitrating. Paper made from the pulps of sulphite and sulphate processes is capable of yielding a very soluble pyroxyline. It can be nitrated at high temperatures and still yield good results. Tissue paper made from flax fibre is also used after being cut into squares.

Mowbray (U.S.P., No. 443, 105, 3rd December 1890) says that a pure cotton tissue paper less than 1/500 inch in thickness, thin as it is, takes on a glutinous or colloid surface, and thus requires some thirty minutes to enable the nitration to take place. With a thicker paper only the surface would be nitrated. He therefore uses a fibre that has been saturated with a solution of nitrate of soda, and afterwards dried slowly, claiming that the salt crystallises in the fibre, or enters by the action termed osmose, and opens up the fibre to the action of the acid. This process would only be useful when the cotton is to be nitrated at a low temperature. At a high temperature it would be unnecessary.

Dietz and Wayne (U.S.P., No. 133, 969) use ramie, rheca, or China grass for producing a soluble pyroxyline. That made from ramie is always of uniform strength and solubility, and requires a smaller quantity of solvent to dissolve it than that made from cotton. Mr Field's experience, however, is entirely contrary to this statement. Such is the influence of the physical form of the fibre on the process of nitration, that when flax fibre and cotton fibre are nitrated with acid mixtures of exactly the same strength, and at the same temperature, the solution of the first is glutinous or thick, and the second fluid or thin. By simply nitrating at a higher temperature than the cotton, the flax will yield a pyroxyline giving an equally fluid collodion.

The presence of chlorine in the fibre must be carefully avoided, as such a fibre will yield an acid product which cannot be washed neutral. The fibre must be dry before nitration; and this is best done, according to Mr Field, by using the form of drier used in drying wool.

~Nitration of the Fibre.~ — Mixed cotton and flax fibre in the form of paper, from 2/1000 to 3/1000 inch thick, and cut into 1-inch squares, is nitrated by the Celluloid Manufacturing Company, and the same paper, left in long strips, 1 inch wide, is used for nitration by the Xylonite Manufacturing Company, of North Adams, Mass. (U.S.A.).

The Celluloid Company introduce the cut paper into the mixed acids by means of a hollow, rapidly revolving tube, flared at the lower end, and immersed in the mixed acids. The centrifugal force of the revolving tube throws the paper towards the sides of the vessel, leaving the centre of the vessel ready for fresh paper.

The Xylonite Company simply cut the paper into long strips, and introduce it into the mixed acids by means of forks. The arrangement used by this Company for holding the mixed acids is a cylindrical vessel divided into a number of sections, the whole revolving like a turntable, thus allowing the workman to nitrate successively each lot of paper at a given point. This Company did not remove the acid from the paper after its immersion, but plunged it immediately into the water, thus losing a large proportion of the waste acid. The Celluloid Company, by using the paper in smaller pieces, and more paper to a pound of acid, and wringing the mixed acid from the paper before immersion in water, had a better process of nitration.

Other manufacturers use earthenware vessels, and glass or steel rods, hooked at one end, having small pieces of rubber hose pulled over the other end to prevent the hand from slipping. The form of vessel in general use is that given in Fig. 23. It is large enough to nitrate 1 lb. of cotton at a time. The hook at one end of the rod enables the workman to pull the pyroxyline apart, and thus ensures saturation of the fibre. In the winter the room in which the nitrating is done must be kept at a temperature of about 70° F. in order to secure equality in the batches.

[Illustration: FIG. 23.—VESSEL FOR NITRATING COTTON OR PAPER.]

The nitrating apparatus of White and Schupphaus (U.S.P., No. 418, 237, 89) Mr Field considers to be both novel and excellent. The cage (Fig. 24), with its central perforated cylinder (Fig. 25), is intended to ensure the rapid and perfect saturation of the tissue paper used for nitrating. The patentees say that no stirring is required with their apparatus. This, says Mr Field, might be true when paper is used, or even cotton, when the temperature of nitration is from 30° to 35° C., but would not be true if the temperature were raised to 50° to 55° C. The process is as follows:— The paper is nitrated in the cage (Fig. 25), the bottom of which is formed by the flanged plate C, fastened to the bottom of the internal cylinder B. After nitration the cage is carried to a wringer, which forms the basket, and the acids removed. Finally, the cage is taken to a plunge tank, where the paper is removed from the cage by simply pulling out the central perforated cylinder B. Fig. 26 shows the nitrating pot, with its automatic cover. The plunge tank is shown in plan and section in Figs. 28 and 29. This apparatus is suitable for the nitration of cotton fibre in bulk at high or low temperatures. Other methods that have been patented are Mowbray's (U.S.P., No. 434, 287), in which it is proposed to nitrate paper in continuous lengths, and Hyatt's (U.S.P., No. 210, 611).

[Illustration: FIG. 24.—CENTRAL PERFORATED CYLINDER.]

[Illustration: FIG. 25.—THE CAGE. WHITE AND SCHUPPHAUS' NITRATING APPARATUS.]

[Illustration: FIG. 26.—CELLULOID NITRATING POT.]

[Illustration: FIG. 27.—ANOTHER VIEW.]

[Illustration: FIGS. 28, 29.—PLUNGE TANK, IN PLAN AND SECTION.]

~The Acid Mixture.~—Various formulæ have been published for producing soluble nitro-cellulose. In many instances, although the observations were correct for the single experiment, a dozen experiments would have produced a dozen different products. The composition of the acids used depends upon the substance to be nitrated, and the temperature at which the nitration will be worked. Practically there are three formulæ in general use—the one used by the celluloid manufacturers; another in which the cotton is nitrated at high temperatures; and a third in which the temperature of the immersion is low, and the time of nitration about six hours. Of the three, the best method is the last one, or the one in which the cotton is immersed at a low temperature, and then the reaction allowed to proceed in pots holding from 5 to 10 lbs. of cotton. The formula used by the celluloid manufacturers for the production of the low form of nitrated product which they use is:—

Sulphuric acid 66 parts by weight.
Nitric acid 17 " "
Water 17 " "

Temperature of immersion, 30° C. Time, twenty to thirty minutes.

The cellulose is used in the form of tissue paper 2/1000 inch thick, 1 lb. to 100 of acid mixture. The nitro-cellulose produced by this formula is very insoluble in the compound ethers and other solvents of pyroxyline, and is seemingly only converted or gelatinised by the action of the solvent. The next formula produces a mixture of tetra- and penta-nitro- celluloses hardly soluble in methyl-alcohol (free from acetone), but very soluble in anhydrous compound ethers, ketones, and aldehydes:—

Nitric acid, sp. gr. 1.435 8 lbs.
Sulphuric acid, sp. gr. 1.83 15-3/4 lbs.
Cotton 14 oz.

Temperature of nitration, 60° C. Time of immersion, forty-five minutes.

The 60° of temperature is developed by mixing the acids together. The cotton is allowed to remain in the acid until it feels "short" to the rod.

The following table, due to Mr W.D. Field, shows very plainly the great variation in the time of the immersion and the temperature by seemingly very slight causes. It extends over fourteen working days, during which time it rained four days. The formula used is that given above, except that the specific gravity of the nitric acid is somewhat lower. The product obtained differs only from that produced by using nitric acid of specific gravity 1.43 in being soluble in methyl-alcohol. From 30 to 35 lbs. of pyroxyline were produced in each of the fourteen days.

A careful examination of this table will prove very instructive. The increase in yield varies from 31 per cent. to nothing, and the loss runs as high as 10 per cent., yet care was taken to make the product uniform in quality. On the days it rained there was a loss, with the exception of the fourth day, when there was neither a loss nor a gain. On the days it was partly clear, as just before or after rain, the table shows a loss in product. We can explain this fact by reason of the moisture-absorbing qualities of the cotton. On the rainy days it would absorb the moisture from the air until, when immersed in the acids, they were weakened, and the fibre dissolved more or less in weakened acid, producing what is known as "burning" in the batch. It will also be noticed that on days which show a loss, the time of the immersion was correspondingly short, as on the a loss, the time of the immersion was correspondingly short, as on the tenth, twelfth, and seventh days.

						Specific Gravity.	Time.		

	H$_2$SO$_4$.	HNO$_3$.	Hours.	Minutes.	Hours.	Minutes.	Temp., Deg. C. From	To	Increase.	Percentage Loss.
1. Clear	1.838	1.4249	...	20	4	...	57°	62°	31	...
2. "	1.837	1.4249	...	20	2	...	60°	62°	18	...
3. Cloudy	1.837	1.4226	...	45	2	...	60°	62°	7	...
4. Rain	1.837	1.420	...	20	1	20	60°	63°	0	0
5. Clear	1.8377	1.42	1	15	2	...	58°	62°	15	...
6. Rainy	1.8391	1.422	...	35	1	40	58°	62°	...	2
7. Cloudy	1.835	1.4226	...	20	...	35	62°	65°	...	10
8. Clear	1.835	1.422	...	35	1	10	60°	62°	5	...
9. Partly Clear	1.824	1.4271	...	20	1	...	50°	60°	...	3
10. "	1.83	1.4271	...	10	...	25	58°	60°	...	10
11. Cloudy	1.832	1.425	...	10	...	50	58°	60°	8	...
12. Rainy	1.822	1.425	...	10	...	20	58°	60°	...	10
13. Partly CLear	1.8378	1.4257	...	60	1	40	50°	58°	20	...
14. Cloudy	1.837	1.4257	1	56	4	40	50°	60°	16	...

The lesson this table teaches is, that it is almost impossible to nitrate cellulose in small quantities, and get uniform results, when the nitration is carried on at high temperatures. As regards the solubility of pyroxyline, Parks found that nitro-benzene, aniline, glacial acetic acid, and camphor, dissolved in the more volatile solvents methyl-alcohol and alcohol-ether, were much the best solvents for producing a plastic, as they are less volatile, and develop greater solvent action under the influence of heat. Nitro-benzene gives a solution that is granular; it seems to merely convert the pyroxyline, and not to dissolve it; but on the addition of alcohol, a solution is at once obtained, and the granular appearance disappears, and the solution becomes homogeneous. The acid mixture and the method of nitrating have much to do with the action of the various solvents, so also has the presence of water.

Dr Schupphaus found that propyl and isobutyl alcohols with camphor were active solvents, and the ketones, palmitone, and stearone in alcohol solution, also alpha- and beta-naphthol, with alcohol and anthraquinone (diphenylene diketone) in alcoholic solution, and also iso-valeric aldehyde and its derivatives, amyliden-dimethyl and amyliden-diethyl ethers.

August Sayer (U.S.P., No. 470,451) finds diethyl-ketone, dibutyl-ketone, di-pentyl-ketone, and the mixed ketones,[A] methyl-ethyl, methyl-propyl, methyl-butyl, methyl-amyl, and ethyl-butyl ketones are active solvents of pyroxyline; and Paget finds that although methyl-amyl oxide is a solvent, that ethyl-amyl oxide is not.

[Footnote A: Ketones are derived from the fatty acids by the substitution of the hydroxyl of the latter by a monad positive radical. They thus resemble aldehydes in constitution. The best-known ketone is acetone $CH_3CO.CH_3$. Mixed ketones are obtained by distilling together salts of two different fatty acids. Thus potassic butyrate and potassic acetate form propyl-methyl-ketone—

$$\begin{array}{l} C(C_2H_5)H_2 \\ | \\ CO.CH_3 \end{array}]$$

The solvents of pyroxyline can be divided into general classes— First, those which are solvents without the aid of heat or solution in alcohol; second, those that are solvents when dissolved in alcohol. These solvents are those which also develop a solvent action when heated to their melting point in combination with pyroxyline.

Mr W.D. Field groups the solvents of pyroxyline into classes thus: Two of the monohydric alcohols; compound ethers of the fatty acids with monohydric alcohols, aldehydes; simple and mixed ketones of the fatty acid series. These four classes include the greater number of the solvents of pyroxyline. Those not included are as follows:— Amyl-nitrate and nitrite, methylene-di-methyl ether, ethidene-diethyl ether, amyl-chloracetate, nitro-benzene and di-nitro-benzene, coumarin, camphor, glacial acetic acid, and mono-, di-, and tri-acetin.

Richard Hale uses the following solvent:—Amyl-acetate, 4 volumes; petroleum naphtha, 4 volumes; methyl-alcohol, 2 volumes; pyroxyline, 4 to 5 ounces to the gallon of solvent. Hale used petroleum naphtha to hasten the drying qualities of the varnish, so that it would set on the article to be varnished before it had a chance to run off. It is, however, the non-hygroscopic character of the solvent that makes the varnish successful. This formula is very largely used for the production of pyroxyline varnish, which is used for varnishing pens, pencils, &c., also brass-work and silver-ware.

The body known as oxy-cellulose[A] is formed by the action of nitric acid upon cellulose when boiled with it. The quantity formed is about 30 per cent. of cellulose acted upon. When washed free from acid, it gelatinises. It is then soluble in dilute alkalies, and can be reprecipitated from solution by alcohol, acids, or saline solutions. Messrs Cross and Bevan assign to it the formula $C_{18}H_{26}O_{16}$. It dissolves in concentrated sulphuric acid, and with nitric acid forms a nitro body of the formula $C_{18}H_{23}O_{16}3(NO_{2})$, which is prepared as follows:—The gelatinous oxy-cellulose is washed with strong nitric acid until free from water, and is then diffused through a mixture of equal volumes of strong sulphuric and nitric acids, in which it quickly dissolves. The solution, after standing for about an hour, is poured in a fine stream into a large volume of water, by which the "nitro" body is precipitated as a white flocculent mass. The product, after drying at 110° C., was found upon analysis to contain 6.48 per cent. nitrogen.

[Footnote A: "On the Oxidation of Cellulose," by C.F. Cross and E.J. Bevan, *Jour. Chem. Soc.*, 1883, p. 22.]

MISCELLANEOUS NITRO-EXPLOSIVES.

~Nitro-Starch.~—It is only recently that, by means of the process introduced by the "Actiengesellschaft Dynamit Nobel," it has been possible to make this explosive upon the manufacturing scale. Nitro-starch has been known since 1883, when Braconnot discovered it, and called it xyloidine. Its formula is $C_6H_8O_3(NO_3)_2$, but Dr Otto Mühlhäusen has lately succeeded in preparing higher nitrated compounds, viz.:—

(*a*.) $C_6H_{7-1/2}O_{2-1/2}(NO_3)_{2-1/2}$.

(*b*.) $C_6H_7O_4(NO_3)_3$.

Or doubling the molecule of starch:—

Nitrogen.
i. *Tetra-nitro-starch* $C_{12}H_{16}O_6(ONO_2)_4$ *11.11 per cent.*
ii. *Penta-nitro-starch* $C_{12}H_{15}O_5(ONO_2)_5$ *12.75* "
iii. *Hexa-nitro-starch* $C_{12}H_{14}O_4(ONO_2)_6$ 14.14 "

He regards them as true ethers (esters) of nitric acid. Thus on treatment with sulphuric acid, these compounds yield NO_3H, the residue $O.NO_2$ thus appearing to be replaced by the sulphuric acid residue. On treatment with a solution of ferrous chloride, nitric oxide and "soluble" starch are regenerated. On shaking with sulphuric acid over mercury, all the nitrogen is split off as NO.

Tetra-nitro-starch is prepared upon the large scale as follows:—A quantity of potato-starch is taken and exposed in some suitable desiccating apparatus at a temperature of 100° C. until all the moisture which it contains is completely driven off. It is then reduced to a fine powder by grinding, and dissolved in nitric acid of specific gravity 1.501. The vessel in which this solution is accomplished is made of lead, and must be provided with two jackets, cooled by means of water. It should further be fitted with a screw-agitator, in order to keep the nitric acid circulating freely. The charge of starch is introduced through an opening in the cover of this digesting vessel, and the proportions of acid to starch are 10 kilogrammes of starch to 100 kilos. of acid. The temperature is kept within the limits 20° to 25° C. When the solution of the starch is complete, the liquid is conducted into a precipitating apparatus, which is also provided with a cooling jacket, for the purpose of regulating the temperature. The bottom of this vessel is double and perforated, and here is placed a layer of gun-cotton to act as a filter. This vessel is filled with spent nitro-sulphuric acid obtained as a waste product from the nitro-glycerine manufactory, and the solution of starch in nitric acid is sprayed into it through an injector worked by compressed air, whereby the nitro-starch is thrown down in the form of a fine-grained powdery precipitate.

In order to precipitate 100 kilos. of the acid solution of starch, it is necessary to employ 500 kilos. of spent nitro-sulphuric acid. As it is precipitated the nitro-starch collects on the gun-cotton filter, and the acid liquor is run off through a tap placed beneath the perforated double bottom of the vessel, and of course below the filter pad. The precipitated starch is further cleansed from acid by repeated washings and by pressure, until all trace of acidity has been eliminated, and the substance exhibits a neutral reaction. The next step is to treat the nitro-starch with a 5 per cent. solution of soda, in contact with which it is allowed to stand for at least twenty-four hours. The product is then ground up until a sort of "milk" or emulsion is obtained, and lastly treated with a solution of aniline, so that when pressed into cake, it contains about 33 per cent. of water, and 1 per cent. of aniline.

Dr Mühlhäusen, working on these lines in the laboratory, prepared nitro- starch which contained 10.96 and 11.09 per cent. of nitrogen. When in the state of powder it is snow-white in colour; it becomes electrified when rubbed; it is very stable, and soluble even in the cold in nitro- glycerine. He has also prepared a tetra-nitro-starch containing 10.58 and 10.50 per cent. of nitrogen, by pouring water into a solution of starch in nitric acid which had stood for several days. The substance thus produced in the laboratory had all the properties of that prepared by the other process.

The production of penta-nitro-starch is effected by adding 20 grms. of rice-starch—previously dried at a temperature of 100°C., in order to eliminate all moisture—to a mixture of 100 grms. of nitric acid, specific gravity 1.501, and 300 grms. of sulphuric acid, specific gravity 1.8 (some tetra-nitro-starch is also formed at the same time). After standing in contact with these mixed acids for one hour the starch has undergone a change, and the mass may now be discharged into a large quantity of water, and then washed, first with water, and finally with an aqueous solution of soda. The yield in Dr Mühlhäusen's experiments was 147.5 per cent.

The substance thus formed is now heated with ether-alcohol, the ether is distilled off, and the penta-nitro-starch appears as a precipitate, whilst the tetra-nitro-starch, which is formed simultaneously, remains in solution in the alcohol. As obtained by this process, it contained 12.76 and 12.98 per cent. nitrogen, whilst the soluble tetra-nitro-starch contained 10.45 per cent.

Hexa-nitro-starch is the product chiefly formed when 40 grms. of dry starch are treated with 400 grms. of nitric acid, specific gravity 1.501, and allowed to stand in contact for twenty-four hours; 200 grms. of this mixture are then poured into 600 c.c. of sulphuric acid of 66° B. The result of this manipulation is a white precipitate, which contains 13.52-13.23 and 13.22 per cent. nitrogen; and consists, therefore, of a mixture of penta- and hexa-nitro-starch.

The experiments undertaken with these substances demonstrated that those prepared by precipitating the nitro-starch with strong sulphuric acid were less stable in character or properties than those which were precipitated by water or weak sulphuric acid. Dr Mühlhäusen is of opinion that possibly in the former case a sulpho-group may be formed, which in small quantity may occasion this instability.

The following table shows the behaviour of these substances prepared in different ways and under various conditions:—

	SAMPLES.				
	A.	B.	C.	D.	E.
Ignition-point	175° C.	170° C.	152° C.	121° C.	155° C.
Stability	Stable	Stable	Unstable	Unstable	Unstable
Per cent. of N.	11.02	10.54	12.87	12.59	13.52
96 per cent. alcohol	Sol.	Sol.	Insol.	Insol.	Insol.
Ether	Insol.	Insol.	Insol.	Insol.	Insol.
Ether-alcohol	Sol.	Sol.	Sol.	Sol.	Sol.
Acetic Ether	Sol.	Sol.	Sol.	Sol.	Sol.

These samples were prepared as follows:—

A. From 1 part nitric acid and 2 parts sulphuric acid (containing 70 per
   cent. H_2O).
B. From 1 part nitric acid and water.
C. From 1 part nitric and 3 parts H_2SO_4 (con.).
D. From 1 part nitric and 3.5 parts con. H_2SO_4.
E. From 1 part nitric and 3 parts con. H_2SO_4.

Dr Mühlhäusen is of opinion that these compounds may be turned to practical account in the production of good smokeless powder. He recommends the following proportions and method. Six grms. of nitro-jute and 2 grms. of nitro-starch are mixed together, and moistened with acetic ether. These ingredients are then worked together into a uniform mass, and dried at a temperature ranging between the limits 50° to 60° C. He has himself prepared such a smokeless powder, which proved to contain 11.54 per cent. of nitrogen, and was very stable. Further details of Dr Mühlhäusen's work upon nitro-starch can be found in *Dingler's Polytechnisches Journal*, paper "Die höhren Salpetersäureäther der Stärke," 1892, Band 284, s. 137-143, and a Bibliography up to 1892 in *Arms and Explosives*, December 1892.

M. Berthelot gives the heat of formation of nitro-starch as 812 cals. for 1 grm., and the heat of total combustion as equal to 706.5 cals. for 207 grms., or for 1 grm. 3,413 cals. The heat of decomposition could only be calculated if the products of decomposition were given, but they have not as yet been studied, and the quantity of oxygen contained in the compound is far from being sufficient for its complete combustion. Berthelot and Vieille found the average velocities for nitro-starch powder, density of charge about 1.2, in a tin tube 4 mm. external diameter, to be, in two experiments, 5,222 m. and 5,674 m. In a tin tube 5.5 mm. external diameter, the velocity was 5,815 m., and in lead tube 5,006 m. (density 1.1 to 1.2). The starch powder is hygroscopic, and is insoluble in water and alcohol. When dry it is very explosive, and takes fire at about 350° F. Mr Alfred Nobel has taken out a patent (Eng. Pat. No. 6,560, 88) for the use of nitro-starch. His invention relates to the treatment of nitro-starch and nitro-dextrine, for the purpose of producing an explosive powder, to be used in place of gunpowder. He incorporates these materials with nitro-cellulose, and dissolves the whole in acetone, which is afterwards distilled off. A perfect incorporation of the ingredients is thus brought about.

~Nitro-Jute.~—It is obtained by treating jute with nitric acid. Its properties have been studied by Messrs Cross and Bevan (*Jour. Chem. Soc.*, 1889, 199), and by Mühlhäusen. The latter used for its nitration an acid mixture composed of equal parts of nitric and sulphuric acids, which was allowed to act upon the jute for some time. He found that with long exposure, i.e., from three to four hours in the acids, there was a disintegrating of the fibre-bundles, and the nitration was attended by secondary decomposition and conversion into products soluble in the acid mixture. Cross and Bevan's work upon this subject leads them to conclude that the highest yield of nitrate is represented by an increase of weight of 51 per cent. They give jute the empirical formula $C_{12}H_{18}O_9$ (C = 47 per cent. H = 6 per cent., and O = 47 per cent.), and believe its conversion into a nitro compound to take place thus:—

$C_{12}H_{18}O + 3HNO_3 = C_{12}H_{15}O_6(NO_3)_{3} + 3H_2O.$

This is equivalent to a gain in weight of 44 per cent. for the tri- nitrate, and of 58 per cent. for the tetra-nitrate. The formation of the tetra-nitrate appears to be the limit of nitration of jute-fibre. In other words, if we represent the ligno-cellulose molecule by a C_{12} formula, it will contain four hydroxyl (OH) groups, or two less than cellulose similarly represented. The following are their nitration results:—

Acids used.—I. HNO_3 sp. gr. 1.43, and H_2SO_4 = 1.84 equal parts.
  II. 1 vol. HNO_3(1.5), 1 vol. H_2SO_4(1.84).
  III. 1 vol. HNO_3(1.5), 75 vols. H_2SO_4(1.84).

I. = 144.4; II. = 153.3; III. = 154.4 grms.; 100 grms. of fibre being used in all three cases.

Duration of exposure, thirty minutes at 18° C.

The nitrogen was determined in the products, and equalled 10.5 per cent. Theory for $C_{12}H_{15}O_{6}(NO_{3})_{3}$ = *9.5 per cent. and for* $C_{12}H_{15}O_{6}(NO_{3})_{4}$ = 11.5 per cent. These nitrates resemble those of cellulose, and are in all essential points nitrates of ligno-cellulose.

Mühlhäusen obtained a much lower yield, and probably, as pointed out by Cross and Bevan, a secondary decomposition took place, and his products, therefore, probably approximate to the derivatives of cellulose rather than to those of ligno-cellulose, the more oxidisable, non-cellulose, or lignone constituents having been decomposed. In fact, he regards his product as cellulose penta-nitrate ($C_{12}H_{16}O_{5}(ONO_{2})_{5}$). The *Chemiker Zeitung*, xxi., p. 163, contains a further paper by Mühlhäusen on the explosive nitro-jute. After purifying the jute-fibre by boiling it with a 1 per cent. solution of sodium carbonate, and washing with water, he treated 1 part of the purified jute with 15 parts of nitro-sulphuric acid, and obtained the following results with different proportions of nitric to sulphuric acids:—

Yield Ignition Nitrogen.
per cent. Point.

Experiment I.— 1. HNO_{3} 1. $H_{2}SO_{4}$ 129.5 170° C. 11.96%
" II. " 2. " 132.2 167° C. 12.15%
" III. " 3. " 135.8 169° C. 11.91%

An experiment made with fine carded jute and the same mixture of acids as in No. II. gave 145.4 per cent. nitro-jute, which ignited at 192° C., and contained 12 per cent. nitrogen. This explosive is not at present manufactured upon the large scale, and Messrs Cross and Bevan are of opinion that there is no very obvious advantage in the use of lignified textile fibre as raw materials for explosive nitrates, seeing that a large number of raw materials containing cellulose (chiefly as cotton) can be obtained at a cheaper rate, and yield also 150 to 170 per cent. of explosive material when nitrated, and are in many ways superior to the products obtained hitherto from jute.

~Nitro-mannite~ is formed by the action of nitric acid on mannite, a hex-acid alcohol closely related to sugar. It occurs abundantly in manna, which is the partly dried sap of the manna-ash (*Fraxinus ornus*). It is formed in the lactic acid fermentation of sugar, and by the action of nascent hydrogen on glucose and cellulose, or on invert sugar. Its formula is $C_6H_8(OH)_6$ *and that of nitro-mannite* $C_6H_8(NO_3)_6$. Mannite crystallises in needles or rhombic prisms, which are soluble in water and alcohol, and have a sweet taste. Nitro-mannite forms white needle-shaped crystals, insoluble in water, but soluble in ether or alcohol. When rapidly heated, they ignite at about 374° F., and explode at about 590° F. It is more susceptible to friction and percussion than nitro-glycerine, and unless pure it is liable to spontaneous decomposition. It is considered as the nitric ether of the hexatomic alcohol mannite. It is formed by the action of a mixture of nitric and sulphuric acids upon mannite—

$$C_6H_8(OH)_6 + 6HNO_3 = C_6H_8(NO_3)_6 + 6H_2O.$$

Its products of explosion are as shown in the following equation:—

$$C_6H_8(OH)_6 = 6CO_2 + 4H_2O + 3N_2 + O_2.$$

Its percentage composition is as follows:—Carbon, 15.9 per cent.; hydrogen, 1.8 per cent.; nitrogen, 18.6 per cent.; and oxygen, 63.7 per cent. Its melting point is 112 to 113° C., and it solidifies at 93°. When carefully prepared and purified by recrystallisation from alcohol, and kept protected from sunlight, it can be kept for several years without alteration.

Nitro-mannite is more dangerous than nitro-glycerine, as it is more sensitive to shock. It is intermediate in its shattering properties between nitro-glycerine and fulminate of mercury. It explodes by the shock of copper on iron or copper, and even of porcelain on porcelain, provided the latter shock be violent. Its heat of formation from its elements is +156.1 calories. It is not manufactured upon the commercial scale.

Besides the nitro compounds already described, there are many others, but they are of little importance, and are none of them made upon the large scale. Among such substances are *nitro-coal*, which is made by the action of nitric acid on coal; *nitro-colle*, a product which results from the action of nitric acid on isinglass or gelatine, soaked in water. It is then treated with the usual acids.

Another method is to place strong glue in cold water until it has absorbed the maximum amount of the latter. The mixture is solidified by the addition of nitric acid, nitrated in the usual way, and well washed. Abel's *Glyoxiline* is only nitrated gun-cotton impregnated with nitro- glycerine. Nitro-lignine is only nitro-cellulose made from wood instead of cotton; and nitro-straw is also only nitro-cellulose. The explosive known as *Keil's Explosive* contains nitro-glucose. Nitro-molasses, which is a liquid product, has also been proposed, and nitro-saccharose, the product obtained by the nitration of sugar. It is a white, sandy, explosive substance, soluble in alcohol and ether. When made from cane sugar, it does not crystallise; but if made from milk sugar, it does. It has been used in percussion caps, being stronger and quicker than nitro-glycerine. It is, however, very sensitive and very hygroscopic, and very prone to decomposition. Nitro-tar, made from crude tar-oil, by nitration with nitric acid of a specific gravity of 1.53 to 1.54. Nitro-toluol is used, mixed with nitro-glycerine. This list, however, does not exhaust the various substances that have been nitrated and proposed as explosives. Even such unlikely substances as horse dung have been experimented with. None of them are very much used, and very few of them are made upon the manufacturing scale.

CHAPTER IV.

DYNAMITE AND GELATINES.

Kieselguhr Dynamite—Classification of Dynamites—Properties and Efficiency of Ordinary Dynamite—Other Forms of Dynamite—Gelatine and
Gelatine Dynamites, Suitable Gun-Cotton for, and Treatment of—Other
Materials used—Composition of Gelignite—Blasting Gelatine—Gelatine
Dynamite—Absorbing Materials—Wood Pulp—Potassium Nitrate, &c.—
Manufacture and Apparatus used, and Properties of Gelatine Dynamites—
Cordite—Composition and Manufacture.

~Dynamite.~—Dynamite consists of nitro-glycerine either absorbed by some porous material, or mixed with some other substance or substances which are either explosives or merely inert materials. Among the porous substances used is kieselguhr, a silicious earth which consists chiefly of the skeletons of various species of diatoms. This earth occurs in beds chiefly in Hanover, Sweden, and Scotland. The best quality for the purpose of manufacturing dynamite is that which contains the largest quantity of the long tubular *bacillariæ*, and less of the round and lancet-shaped forms, such as *pleurosigmata* and *diclyochæ*, as the tube-shaped diatoms absorb the nitro-glycerine better, and it becomes packed into the centre of the silicious skeleton of the diatoms, the skeleton acting as a kind of tamping, and increasing the intensity of the explosion.

Dynamites are classified by the late Colonel Cundill, R.A., in his "Dictionary of Explosives" as follows:—

1. Dynamites with an inert base, acting merely as an absorbent.

2. Dynamites with an active base, i.e., an explosive base. No. 2 may be again divided into three minor classes, which contain as base—

(*a*.) Charcoal.

(*b*.) Gunpowder or other nitrate, or chlorate mixture.

(*c*.) Gun-cotton or other nitro compound (nitro-benzol, &c.).

The first of these, viz., charcoal, was one of the first absorbents for nitro-glycerine ever used; the second is represented by the well-known Atlas powder; and the last includes the well-known and largely used gelatine compounds, viz., gelignite and gelatine dynamite, and also tonite No. 3, &c.

In the year 1867 Nobel produced dynamite by absorbing the nitro-glycerine in an inert substance, forming a plastic mass. In his patent he says: "This invention relates to the use of nitro-glycerine in an altered condition, which renders it far more practical and safe for use. The altered condition of the nitro-glycerine is effected by causing it to be absorbed in porous unexplosive substances, such as charcoal, silica, paper, or similar materials, whereby it is converted into a powder, which I call dynamite, or Nobel's safety powder. By the absorption of the nitro- glycerine in some porous substance it acquires the property of being in a high degree insensible to shocks, and it can also be burned over a fire without exploding."

Ordinary dynamite consists of a mixture of 75 per cent. of nitro-glycerine and 25 per cent. of kieselguhr. The guhr as imported (Messrs A. Haake & Co. are the chief importers) contains from 20 to 30 per cent. of water and organic matter. The water may be very easily estimated by drying a weighed quantity in a platinum crucible at 100° C. for some time and re-weighing, and the organic matter by igniting the residue strongly over a Bunsen burner. Before the guhr can be used for making dynamite it must be calcined, in order not only to get rid of moisture, but also the organic matter.

A good guhr should absorb four times its weight of nitro-glycerine, and should then form a comparatively dry mixture. It should be pale pink, red brown, or white. The pink is generally preferred, and it should be as free as possible from grit of all kinds, quartz particles, &c., and should have a smooth feeling when rubbed between the finger and thumb, and should show a large quantity of diatoms when viewed under the microscope. The following was the analysis of a dried sample of kieselguhr:—Silica, 94.30; magnesia, 2.10; oxide of iron and alumina, 1.3; organic matter, 0.40; moisture, 1.90 per cent.

The guhr is generally dried in a reverberatory muffle furnace. It is spread out on the bottom to the thickness of 3 or 4 inches, and should every now and then be turned over and raked about with an iron rabble or hoe. The temperature should be sufficiently high to make the guhr red hot, or the organic matter will not be burnt off. The time occupied in calcining will depend of course upon the quality of the guhr being operated upon. Those containing a high percentage of water and organic matter will of course take longer than those that do not. A sample of the calcined guhr should not contain more than 0.5 per cent. of moisture and organic matter together.

After the guhr is dry it requires to be sifted and crushed. The crushing is done by passing it between iron rollers fixed at the bottom of a cone or hopper, and revolving at a moderate speed. Beneath the rollers a fine sieve should be placed, through which the guhr must be made to pass.

The kieselguhr having been dried, crushed, and sifted, should be packed away in bags, and care should be taken that it does not again absorb moisture, as if it contains anything above about five-tenths per cent. of water it will cause the dynamite made with it to exude. The guhr thus prepared is taken up to the danger area, and mixed with nitro-glycerine. The nitro-glycerine used should be quite free from water, and clear, and should have been standing for a day or two in the precipitating house. The guhr and nitro-glycerine are mixed in lead tanks (about 1-1/2 foot deep, and 2 to 3 feet long), in the proportions of 75 of the nitro-glycerine to 25 of the guhr, unless the guhr is found to be too absorbent, which will cause the dynamite to be too dry and to crumble. In this case a small quantity of barium sulphate, say about 1 per cent., should be added to the guhr. This will lessen its absorbing powers, or a highly absorptive sample of guhr may be mixed with one of less absorptive power, in the proportions found by experiment to be the best suited to make a fairly moist dynamite, but one that will not exude.

The mixing itself is generally performed in a separate house. In a series of lead-lined tanks the guhr is weighed, placed in a tank, and the nitro- glycerine poured on to it. The nitro-glycerine may be weighed out in indiarubber buckets. The whole is then mixed by hand, and well rubbed between the hands, and afterwards passed through a sieve. At this stage the dynamite should be dry and powdery, and of a uniform colour.

It is now ready to be made up into cartridges, and should be taken over to the cartridge huts. These are small buildings surrounded with mounds, and contain a single cartridge machine. Each hut requires three girls—one to work the press, and two to wrap up the cartridges. The cartridge press consists of a short cylinder of the diameter of the cartridge that it is intended to make. Into this cylinder a piston, pointed with ivory or lignum vitæ wood, works up and down from a spring worked by a lever. Round the upper edge of the cylinder is fastened a canvas bag, into which the powdery dynamite is placed by means of a wooden scoop, and the descending piston forces the dynamite down the cylinder and out of the open end, where the compressed dynamite can be broken off at convenient lengths. The whole machine should be made of gun-metal, and should be upright against the wall of the building. The two girls, who sit at tables placed on each side of the press, wrap the cartridges in parchment paper. From these huts the cartridges are collected by boys every ten minutes or a quarter of an hour, and taken to the packing room, where they are packed in 5-lb. cardboard boxes, which are then further packed in deal boxes lined with indiarubber, and fastened down air tight. The wooden lids are then nailed down with brass or zinc nails, and a label pasted on the outside giving the weight and description of the contents. The boxes should then be removed to the magazines. It is well to take a certain number of cartridges from the packing house at different times during the day, say three or four samples, and to test them by the heat test. A sample cut from a cartridge, about 1 inch long, should be placed under a glass shade, together with water (a large desiccator, in fact), and left for some days. A good dynamite should not, under these conditions, show any signs of exudation, even after weeks.[A]

[Footnote A: For analysis of dynamite, see chapter on "Analysis," and author's article in *Chem. News*, 23rd September 1892.]

~Properties of Kieselguhr Dynamite.~ — One cubic foot of dynamite weighs 76 lbs. 4 oz. The specific gravity of 75 per cent. dynamite is, however, 1.50. It is a red or grey colour, and rather greasy to the touch. It is much less sensitive to shock than nitro-glycerine, but explodes occasionally with the shock of a rifle bullet, or when struck. The addition of a few per cent. of camphor will considerably diminish its explosive qualities to such an extent that it can be made non-explosive except to a very strong fulminate detonator. The direct contact of water disintegrates dynamite, separating the nitro-glycerine, hence great caution is necessary in using it in wet places. It freezes at about 40° Fahr. (4° C.), and remains frozen at temperatures considerably exceeding that point. When frozen, it is comparatively useless as an explosive agent, and must be thawed with care. This is best done by placing the cartridges in a warming pan, which consists of a tin can, with double sides and bottom, into which hot water (130° Fahr.) can be poured. The dynamite will require to be left in for some considerable time before it becomes soft. On no account must it be placed on a hot stove or near a fire, as many serious accidents have occurred in this way.

Frozen dynamite is a hard mass, with altered properties, and requires 1.5 grm. of fulminate instead of 0.5 grm. to explode it. Thawing may also cause exudation of the nitro-glycerine, which is much more sensitive to shock, and if accidentally struck with an iron tool, may explode. It is a dangerous thing to cut a frozen cartridge with a knife. Ramming is even more dangerous; in fact it is not only dangerous, but wasteful, to use dynamite when in a frozen state.

Dynamite explodes at a temperature of 360° Fahr., and is very sensitive to friction when hot. In hot countries it should never be exposed to the rays of the sun. It should, however, not be kept in a damp or moist place, as this is liable to cause exudation. Sunlight, if direct, can cause a slow decomposition, as with all nitro and nitric compounds. Electric sparks ignite, without exploding it, at least when operating in the open air.

Dynamite, when made with neutral nitro-glycerine, appears to keep indefinitely. Sodium or calcium carbonate to the extent of 1 per cent. is often added to dynamite to ensure its being neutral. If it has commenced to undergo change, however, it rapidly becomes acid, and sometimes explodes spontaneously, especially if contained in resisting envelopes. Nevertheless, neutral and well-made dynamite has been kept for years in a magazine without loss of its explosive force. If water is brought into contact with it, the nitro-glycerine is gradually displaced from the silica (guhr). This action tends to render all wet dynamite dangerous.

It has been observed that a dynamite made with wood sawdust can be moistened and then dried without marked alteration, and from 15 to 20 per cent. of water may be added to cellulose dynamite without depriving it of the power of exploding by strong detonator (this is similar to wet gun-cotton). It is, however, rendered much less sensitive to shock. With regard to the power of No. 1 dynamite, experiments made in lead cylinders give the relative value of No. 1 dynamite, 1.0; blasting gelatine, 1.4; and nitro-glycerine, 1.4. The heat liberated by the sudden explosion of dynamite is the same as its heat of combustion,[A] and proportionate to the weight of nitro-glycerine contained in the mixture. The gases formed are carbonic acid, water, nitrogen, and oxygen.

[Footnote A: Berthelot, "Explosives and their Power."]

The "explosive wave" (of Berthelot) for dynamite is about 5,000 metres per second. At this rate the explosion of a cartridge a foot long would only occupy 1/24000 part of a second, while a ton of dynamite cartridges about 7/8 diameter, laid end to end, and measuring one mile in length, would be exploded in one-quarter of a second by detonating a cartridge at either end.[A] Mr C. Napier Hake, F.I.C., the Inspector of Explosives for the Victorian Government, in his paper, "Notes on Explosives," says: "The theoretical efficiency of an explosive cannot in practice be realised in useful work for several reasons, as for instance in blasting rock—

"1. Incomplete combustion.

"2. Compression and chemical changes induced in surrounding material.

"3. Energy expended in cracking and heating of the material which is not displaced.

"4. The escape of gas through the blast-hole and the fissures caused by the explosion.

"The useful work consists partly in displacing the shattered masses. The proportion of useful work obtainable has been variously estimated at from 14 to 33 per cent. of the theoretical maximum potential."

[Footnote A: C.N. Hake, "Notes on Explosives," *Jour. Soc. Chem. Ind.*, 1889.]

Among the various forms of dynamite that are manufactured is carbo- dynamite, the invention of Messrs Walter F. Reid and W.D. Borland. The base is nitro-glycerine, and the absorbent is carbon in the form of burnt cork. It is as cheap as ordinary dynamite, and has greater explosive force, seeing that 90 per cent. of the mixture is pure nitro-glycerine, and the absorbent itself is highly combustible. It is also claimed that if this dynamite becomes wet, no exudation takes place.

Atlas powder is a dynamite, chiefly manufactured in America at the Repanno Chemical Works, Philadelphia. It is a composition of nitro-glycerine, wood-pulp, nitrate of soda, and carbonate of magnesia. This was the explosive used in the outrages committed in London, by the so-called "dynamiters." Different varieties contain from 20 to 75 per cent. of nitro-glycerine.

The Rhenish dynamite, considerably used in the mines of Cornwall, is composed of 70 parts of a solution of 2 to 3 per cent. of naphthalene in nitro-glycerine, 3 parts of chalk, 7 parts of sulphate of barium, and 20 of kieselguhr.

Kieselguhr dynamites are being largely given up in favour of gelatine explosives. The late Colonel Cundill, in his "Dictionary of Explosives," gives a list of about 125 kinds of dynamites. Many of these, however, are not manufactured. Among the best known after the ordinary No. 1 dynamite are forcite, ammonia dynamite, lithofracteur, rendock, Atlas powder, giant powder, and the various explosive gelatines. They all contain nitro- glycerine, mixed with a variety of other substances, such as absorbent earths, wood-pulp, nitro-cotton, carbon in some form or other, nitro- benzol, paraffin, sulphur, nitrates, or chlorates, &c. &c.

~Blasting Gelatine and Gelatine Dynamite.~ — The gelatine explosives chiefly in use are known under the names of blasting gelatine, gelatine dynamite, and gelignite. They all consist of the variety of nitro- cellulose known as collodion-cotton, i.e., a mixture of the penta- and tetra-nitrates dissolved in nitro-glycerine, and made up with various proportions of wood-pulp, and some nitrate, or other material of a similar nature. As the gun-cotton contains too little oxygen for complete combustion, and the nitro-glycerine an excess, a mixture of the two substances is very beneficial.

Blasting gelatine consists of collodion-cotton and nitro-glycerine without any other substance, and was patented by Mr Alfred Nobel in 1875. It is a clear, semi-transparent, jelly-like substance, of a specific gravity of 1.5 to 1.55, slightly elastic, resembling indiarubber, and generally consists of 92 per cent. to 93 per cent. of nitro-glycerine, and 7 to 8 per cent. of nitro-cotton. The cotton from which it is made should be of good quality. The following is the analysis of a sample of nitro-cellulose which made very good gelatine:-

Soluble cotton 99.118 per cent.
Gun-cotton 0.642 "
Non-nitrated cotton 0.240 "
Nitrogen 11.64 "
Total ash 0.25 "

The soluble cotton, which is a mixture of the tetra- and penta-nitrates, is soluble in ether-alcohol, and also in nitro-glycerine, and many other solvents, whereas the hexa-nitrate (gun-cotton), $C_{12}H_{14}O_{4}(ONO_{2})_{6}$, is not soluble in the above liquids, although it is soluble in acetone or acetic ether. It is very essential, therefore, that the nitro-cotton used in the manufacture of the gelatine explosives should be as free as possible from gun-cotton, otherwise little lumps of undissolved nitro-cotton will be left in the finished gelatine. The non-nitrated or unconverted cotton should also be very low, in fact considerably under 1/2 per cent.

The nitro-cotton and the nitro-glycerine used should always be tested before use by the heat test, because if they do not separately stand this test, it cannot be expected that the gelatine made from them will do so. It often occurs, however, that although both the ingredients stand this test separately before being mixed, that after the process of manufacture one or other or both fail to do so.

The nitro-cotton most suitable for gelatine making is that which has been finely pulped. If it is not already fine enough, it must be passed through a fine brass wire sieve. It will be found that it requires to be rubbed through by hand, and will not go through at all if in the least degree damp. It is better, therefore, to dry it first. The percentage of nitrogen in the nitrated cotton should be over 11 per cent. It should be as free as possible from sand or grit, and should give but little ash upon ignition, not more than 0.25 per cent. The cotton, which is generally packed wet in zinc-lined wooden boxes, will require to be dried, as it is very essential indeed that none of the materials used in the manufacture of gelatine should contain more than the slightest trace of water. If they do, the gelatine subsequently made from them will most certainly exude, and become dangerous and comparatively valueless. It will also be much more difficult to make the nitro-cotton dissolve in the nitro-glycerine if either contains water.

In order to find out how long any sample of cotton requires to be dried, a sample should be taken from the centre of several boxes, well mixed, and about 1,000 grms. spread out on a paper tray, weighed, and the whole then placed in the water oven at 100° C., and dried for an hour or so, and again weighed, and the percentage of moisture calculated from the loss in weight. This will be a guide to the time that the cotton will probably require to be in the drying house. Samples generally contain from 20 to 30 per cent. of water. After drying for a period of forty-eight hours, a sample should be again dried in the oven at 100° C., and the moisture determined, and so on at intervals until the bulk of the cotton is found to be dry, i.e., to contain from 0.25 to 0.5 per cent. of moisture. It is then ready to be sifted. During the process of removing to the sifting house and the sifting itself, the cotton should be exposed to the air as little as possible, as dry nitro-cotton absorbs as much as 2 per cent. of moisture from the air at ordinary temperatures and average dryness.

The drying house usually consists of a wooden building, the inside of which is fitted with shelves, or rather framework to contain drawers, made of wood, with brass or copper wire netting bottoms. A current of hot air is made to pass through the shelves and over the surface of the cotton, which is spread out upon them to the depth of about 2 inches. This current of air can be obtained in any way that may be found convenient, such as by means of a fan or Root's blower, the air being passed over hot bricks, or hot-water pipes before entering the building. The cotton should also be occasionally turned over by hand in order that a fresh surface may be continually exposed to the action of the hot air. The building itself may be heated by means of hot-water pipes, but on no account should any of the pipes be exposed. They should all be most carefully covered over with wood-work, because when the dry nitro-cotton is moved, as in turning it over, very fine particles get into the air, and gradually settling on the pipes, window ledges, &c., may become very hot, when the slightest friction might cause explosion. It is on this account that this house should be very carefully swept out every day. It is also very desirable that the floor of this house should be covered with oilcloth or linoleum, as being soft, it lessens the friction.

List shoes should always be worn in this building, and a thermometer hung up somewhere about the centre of the house, and one should also be kept in one of the trays to give the temperature of the cotton, especially the bottom of the trays. The one nearest to the hot air inlet should be selected. If the temperature of the house is kept at about 40° C. it will be quite high enough. The building must of course be properly ventilated, and it will be found very useful to have the walls made double, and the intervening space filled with cinders, and the roof covered with felt, as this helps to prevent the loss of heat through radiation, and to preserve a uniform temperature, which is very desirable.

The dry cotton thus obtained, if not already fine enough, should be sifted through a brass sieve, and packed away ready for use in zinc air-tight cases, or in indiarubber bags. The various gelatine compounds, gelignite, gelatine dynamite, and blasting gelatine, are manufactured in exactly the same way. The forms known as gelatine dynamite differ from blasting gelatine in containing certain proportions of wood-pulp and potassium nitrate, &c. The following are analyses of some typical samples of the three compounds:—

	Gelignite.	Gelatine Dynamite.	Blasting Gelatine.
Nitro-glycerine	60.514	71.128	92.94 per cent.
Nitro-cellulose	4.888	7.632	7.06 "
Wood-pulp	7.178	4.259	… "
Potassium nitrate	27.420	16.720	… "
Water	…	0.261	… "

The gelignite and gelatine dynamites consist, therefore, of blasting gelatine, thickened up with a mixture of absorbing materials. Although the blasting gelatine is weight for weight more powerful, it is more difficult to make than either of the other two compounds, it being somewhat difficult to make it stand the exudation and melting tests. The higher percentage of nitro-cotton, too, makes it expensive.

When the dry nitro-cotton, which has been carefully weighed out in the proportions necessary either for blasting gelatine or any of the other gelatine explosives, is brought to the gelatine making house, it is placed in a lead-lined trough, and the necessary quantity of pure dry nitro- glycerine poured upon it. The whole is then well stirred up, and kept at a temperature of from 40° to 45° C. It should not be allowed to go much above 40° C.; but higher temperatures may be used if the nitro-cotton is very obstinate,[A] and will not dissolve. Great caution must, however, be observed in this case. The mixture should be constantly worked about by the workman with a wooden paddle for at least half an hour. At a temperature of 40° to 45° the nitro-glycerine acts upon the nitro-cotton and forms a jelly. Without heat the gelatinisation is very imperfect indeed, and at temperatures under 40° C. takes place very slowly.

[Footnote A: Generally due to the nitro-cotton being damp.]

[Illustration: FIG. 30.—WERNER, PFLEIDERER, & PERKINS' MIXING MACHINE.]

The limit of temperature is 50° C. or thereabouts. Beyond this the jelly should never be allowed to go, and to 50° only under exceptional circumstances.

The tank in which the jelly is made is double-lined, in order to allow of the passage of hot water between its inner and outer linings. A series of such tanks are generally built in a wooden framework, and the double linings are made to communicate, so that the hot water can flow from one to the other consecutively. The temperature of the water should be about 60° C. if it is intended to gelatinise at 45° C., and about 80° if at 50° C.; but this point must, of course, be found by experiment for the particular plant used. An arrangement should be made to enable the workman to at once cut off the supply of hot water and pass cold water through the tanks in case the explosive becomes too hot.

[Illustration: FIG. 31. — MR M'ROBERTS' MIXER FOR GELATINE EXPLOSIVES.]

The best way to keep the temperature of the water constant is to have a large tank of water raised upon a platform, some 5 or 6 feet high, outside the building, which is automatically supplied with water, and into which steam is turned. A thermometer stuck through a piece of cork and floated upon the surface of the tank will give the means of regulating the temperature.

When the jelly in the tanks has become semi-transparent and the cotton has entirely dissolved, the mixture should be transferred to the mixing machine. The mixing machines are specially designed for this work, and are built in iron, with steel or bronze kneading- and mixing-blades, according to requirements.

A suitable machine for the purpose is that known as the Nito-Universal Incorporator, shown in Fig. 30, which has been specially constructed by Messrs Werner, Pfleiderer, & Perkins, Ltd., after many years' experience in the mixing of explosive materials, and is now almost exclusively adopted in both Government and private factories. Mr George M'Roberts'[A] mixing machine, however, which is shown in Fig. 31, is still used in some factories for dynamite jelly.

[Footnote A: See *Jour. Soc. Chem. Ind.*, 1890, 267.]

If it is intended to make gelignite, or gelatine dynamite, it is at this point that the proper proportions of wood-pulp[A] and potassium nitrate should be added, and the whole well mixed for at least half an hour, until the various ingredients are thoroughly incorporated.

[Footnote A: Most of the wood-pulp used in England is obtained from pine-trees, but poplar, lime, birch, and beech wood are also used. It is chiefly imported as wood-pulp. The pulp is prepared as follows:—The bark and roots are first removed, and the logs then sawn into boards, from which the knots are removed. The pieces of wood are afterwards put through a machine which breaks them up into small pieces about an inch long, which are then crushed between rollers. These fragments are finally boiled with a solution of sodium bisulphite, under a pressure of about 90 lbs. per square inch, the duration of the boiling being from ten to twelve hours. Sulphurous acid has also been used. Pine-wood yields about 45 per cent. and birch about 40 per cent. of pulp when treated by this process. The pulp is afterwards bleached and washed, &c.

	Birch.	Beech.	Lime.	Pine.	Poplar.
Cellulose	55.52	45.47	53.09	56.99	62.77 per cent.
Resin	1.14	0.41	3.93	0.97	1.37 "
Aqueous extract	2.65	2.47	3.56	1.26	2.88 "
Water	12.48	12.57	10.10	13.87	12.10 "
Lignine	28.21	39.14	29.32	26.91	20.88 "

]

The following analysis of woods is by Dr H. Müller:—These mixing machines can either be turned by hand, or a shaft can be brought into the house and the machine worked by means of a belt at twenty to thirty revolutions per minute. The bearings should be kept constantly greased and examined, and the explosive mixture carefully excluded. When the gelatine mixture has been thoroughly incorporated, and neither particles of nitrate or wood meal can be detected in the mass, it should be transferred to wooden boxes and carried away to the cartridge-making machines to be worked up into cartridges.

[Illustration: FIG. 32.—PLAN OF THE BOX CONTAINING THE EXPLOSIVE, IN

M'ROBERTS' MACHINE.]

The application of heat in the manufacture of the jelly from collodion-cotton and nitro-glycerine is absolutely necessary, unless some other solvent is used besides the nitro-glycerine, such as acetone, acetic ether, methyl, or ethyl-alcohol. (They are all too expensive, with the exception of acetone and methyl-alcohol, for use upon the large scale.) These liquids not only dissolve the nitro-cellulose in the cold, but render the resulting gelatine compound less sensitive to concussion, and reduce its quickness of explosion (as in cordite). They also lower the temperature at which the nitro-glycerine becomes congealed, i.e., they lower the freezing point[A] of the resulting gelatine.

[Footnote A: It has been proposed to mix dynamite with amyl alcohol for this purpose. Di-nitro-mono-chlorhydrine has also been proposed.]

The finished gelatine paste, upon entering the cartridge huts, is at once transferred to the cartridge-making machine, which is very like an ordinary sausage-making machine[A] (Fig. 33). The whole thing must be made of gun-metal or brass, and it consists of a conical case containing a shaft and screw. The revolutions of the shaft cause the thread of the screw to push forward the gelatine introduced by the hopper on the top to the nozzle, the apex of the cone-shaped case, from whence the gelatine issues as a continuous rope. The nozzle is of course of a diameter according to the size of cartridge required.

[Footnote A: G. M'Roberts, *Jour. Soc. Chem. Ind.*, 31st March 1890, p. 266.]

[Illustration: FIG. 33. — CARTRIDGE-MAKING MACHINE FOR GELATINE
EXPLOSIVES.]

The issuing gelatine can of course be cut off at any length. This is best done with a piece of hard wood planed down to a cutting edge, i.e., wedge-shaped. Mr Trench has devised a kind of brass frame, into which the gelatine issuing from the nozzle of the cartridge machine is forced, finding its way along a series of grooves. When the frame is full, a wooden frame, which is hinged to one end of the bottom frame, and fitted with a series of brass knives, is shut down, thereby cutting the gelatine up into lengths of about 4 inches.

It is essential that the cartridge machines should have no metallic contacts inside. The bearing for the screw shaft must be fixed outside the cone containing the gelatine. One of these machines can convert from 5 to 10 cwt. of gelatine into cartridges per diem, depending upon the diameter of the cartridges made.

After being cut up into lengths of about 3 inches, the gelatine is rolled up in cartridge paper. Waterproof paper is generally used. The cartridges are then packed away in cardboard boxes, which are again packed in deal boxes lined with indiarubber, and screwed down air tight, brass screws or zinc or brass nails being used for the purpose. These boxes are sent to the magazines. Before the boxes are fastened down a cartridge or so should be removed and tested by the heat test, the liquefaction test, and the test for liability to exudation. (Appendix, p. 6, Explosives Act, 1875.) A cartridge also should be stored in the magazine in case of any subsequent dispute after the bulk of the material has left the factory.

The object of the liquefaction test is to ensure that the gelatine shall be able to withstand a fairly high temperature (such as it might encounter in a ship's hold) without melting or running together. The test is carried out as follows: — A cylinder of the gelatine dynamite is cut from the cartridge of a length equal to its diameter. The edges must be sharp. This cylinder is to be placed on end on a flat surface (such as paper), and secured by a pin through the centre, and exposed for 144 consecutive hours to a temperature of 85° to 90° F., and during such time the cylinder should not diminish in height by more than one-fourth of an inch, and the cut edges should remain sharp. There should also be no stain of nitroglycerine upon the paper.

The exudation test consists in freezing and thawing the gelatine three times in succession. Under these conditions there should be no exudation of nitro-glycerine. All the materials used in the manufacture of gelatine explosives should be subjected to analytical examination before use, as success largely depends upon the purity of the raw materials. The wood-pulp, for instance, must be examined for acidity.

~Properties of the Gelatine Compounds.~ — Blasting gelatine is generally composed of 93 to 95 parts nitro-glycerine, and 5 to 7 parts of nitro- cellulose, but the relative proportions of explosive base and nitro- glycerine, &c., in the various forms of the gelatine explosives do not always correspond to those necessary for total combustion, either because an incomplete combustion gives rise to a greater volume of gas, or because the rapidity of decomposition and the law of expansion varies according to the relative proportions and the conditions of application. The various additions to blasting gelatine generally have the effect of lowering the strength by reducing the amount of nitro-glycerine, but this is sometimes done in order to change a shattering agent into a propulsive force. If this process be carried too far, we of course lose the advantages due to the presence of nitro-glycerine. There is therefore a limit to these additions.[A]

[Footnote A: Mica is said to increase the rapidity of explosion when mixed with gelatine.]

The homogeneousness and stability of the mixture are of the highest importance. It is highly essential that the nitro-glycerine should be completely absorbed by the substances with which it is mixed, and that it should not subsequently exude when subjected to heat or damp. It is also important that there should be no excess of nitro-glycerine, as this may diminish instead of augment the strength, owing to a difference in the mode of the propagation of the explosive wave in the liquid, and in the mixture. Nitro-glycerine at its freezing point has a tendency to separate from its absorbing material, in fact to exude. When frozen, too, it requires a more powerful detonation to explode it, but it is less sensitive to shock. The specific gravity of blasting gelatine is 1.5 (i.e., nearly equal to that of nitro-glycerol); that of gun-cotton (dry) is 1.0.

Blasting gelatine burns in the air when unconfined without explosion, at least in small quantities and when not previously heated, but it is rather uncertain in this respect. It can be kept at a moderately high temperature (70° C.) without decomposition. At higher temperatures the nitro-glycerine will partially evaporate. When slowly heated, it explodes at 204° C. If, however, it contains as much as 10 per cent. of camphor, it burns without exploding. According to Berthelot,[A] gelatine composed of 91.6 per cent. nitro-glycerine and 8.4 per cent. of nitro-cellulose, which are the proportions corresponding to total combustion, produces by explosion $177CO_2 + 143H_2O + 8N_2$.

[Footnote A: Berthelot, "Explosives and their Powers."]

He takes $C_{24}H_{22}(NO_3H)_9O_{11}$ as the formula of the nitro-cellulose, and $51C_3H_2(NO_3H)_3 + C_{24}H_{22}(NO_3H)_9O_{11}$ as the formula of the gelatine itself, its equivalent weight being 12,360 grms. The heat liberated by its explosion is equal to 19,381 calories, or for 1 kilo. 1,535 calories. Volume of gases reduced temperature equals 8,950 litres. The relative value[A] of blasting gelatine to nitro-glycerine is as 1.4 to 1.45, kieselguhr dynamite being taken as 1.0.

[Footnote A: Roux and Sarran.]

CHAPTER V.

NITRO-BENZOL, ROBURITE, BELLITE, PICRIC ACID, &c.

Explosives derived from Benzene—Toluene and Nitro-Benzene—Di- and
Tri-nitro-Benzene—Roburite: Properties and Manufacture—Bellite: Properties, &c.—Securite—Tonite No. 3.—Nitro-Toluene—Nitro-Naphthalene—Ammonite—Sprengel's Explosives—Picric Acid—
Picrates—Picric Powders—Melinite—Abel's Mixture—Brugère's Powders—
The Fulminates—Composition, Formula, Preparation, Danger of, &c.—
Detonators: Sizes, Composition, Manufacture—Fuses, &c.

~The Explosives derived from Benzene.~—There is a large class of explosives made from the nitrated hydro-carbons—benzene, C_6H_6; toluene, C_7H_8; naphthalene, $C_{10}H_8$; and also from phenol (or carbolic acid), C_6H_5OH. The benzene hydrocarbons are generally colourless liquids, insoluble in water, but soluble in alcohol and ether. They generally distil without decomposition. They burn with a smoky flame, and have an ethereal odour. They are easily nitrated and sulphurated; mono, di, and tri derivatives are readily prepared, according to the strength of the acids used. It is only the H-atoms of the benzene nucleus which enter into reaction.

Benzene was discovered by Faraday in 1825, and detected in coaltar by Hofmann in 1845. It can be obtained from that portion of coaltar which boils at 80° to 85° by fractionating or freezing.[A] The ordinary benzene of commerce contains thiophene (C_4H_4S), from which it may be freed by shaking with sulphuric acid. Its boiling point is 79° C.; specific gravity at 0° equals 0.9. It burns with a luminous smoky flame, and is a good solvent for fats, resins, sulphur, phosphorus, &c. Toluene was discovered in 1837, and is prepared from coal-tar. It boils at 110° C., and is still liquid at 28° C.

[Footnote A: It may be prepared chemically pure by distilling a mixture of benzoic acid and lime.]

The mono-, chloro-, bromo-, and iodo-benzenes are colourless liquids of peculiar odour. Di-chloro-, di-bromo-benzenes, tri- and hexa-chloro- and bromo-benzenes, are also known; and mono-chloro-, $C_6H_4Cl(CH_3)$, and bromo-toluenes, together with di derivatives in the ortho, meta, and para modifications. The nitro-benzenes and toluenes are used as explosives. The following summary is taken from Dr A. Bernthsen's "Organic Chemistry":—

SUMMARY.

_____ | | | $C_6H_5(NO_2)$ Nitro-benzene. Liq. B.Pt. 206° C. | | | | $C_6H_4(NO_2)_2$ *Ortho-, meta-, and para- di-nitro-benzenes.* | | *Solid. M.P. 118°, 90°, and 172° C.* | | | | $C_6H_3(NO_3)_3$ S.-Tri-nitro-benzene. Solid. M.P. 121° C. |

|_____

_____ | | | | $C_6H_4(CH_3)NO_2$ Ortho-, meta-, and para- nitro-toluenes. | | B.P. 218°, 230°, and 234° C, Para compound solid. |

|_____

_____ | | | | $C_6H_3(CH_3)_2NO_2$ Nitro-xylene. Liquid. |

|_____

_____ | | | | $C_6H_2(CH_3)_3NO_2$ Nitro-mesitylene. Solid. |

|_____

_____ | | | | $C_6H_3(CH_3)(NO_2)_2$ Di-nitro-toluenes. |

|_____

_____ | | | | $C_6H_4Cl(NO_2)$ Nitro-chloro-benzenes. | | | | $C_6Br_4(NO_2)_2$ Tetra-bromo-di-nitrobenzene. |

|_____

_____ |

The nitro compounds are mostly pale yellow liquids, which distil unchanged, and volatilise with water vapour, or colourless or pale yellow needles or prisms. Some of them, however, are of an intense yellow colour. Many of them explode upon being heated. They are heavier than water, and insoluble in it, but mostly soluble in alcohol, ether, and glacial acetic acid.

Nitro-benzene, $C_6H_5(NO_2)$, was discovered in 1834 by Mitscherlich. It is a yellow liquid, with a melting point of +3° C. It has an intense odour of bitter almonds. It solidifies in the cold. In di-nitro-benzene, the two nitro groups may be in the meta, ortho, or para position, the meta position being the most general (see fig., page 4). By recrystallising from alcohol, pure meta-di-nitro-benzene may be obtained in long colourless needles. The ortho compound crystallises in tables, and the para in needles. They are both colourless. When toluene is nitrated, the para and ortho are chiefly formed, and a very little of the meta compound.

~Nitro Compounds of Benzene and Toluene.~ — The preparation of the nitro derivatives of the hydrocarbons of the benzene series is very simple. It is only necessary to bring the hydrocarbon into contact with strong nitric acid, when the reaction takes place, and one or more of the hydrogen atoms of the hydrocarbon are replaced by the nitryl group (NO_2). Thus by the action of nitric acid on benzene (or benzol), mono-nitro-benzene is formed:—

$$C_6H_6 + HNO_3 = C_6H_5.NO_2 + H_2O.$$
Mono-nitro-benzene.

By the action of another molecule of nitric acid, the di-nitro-benzene is formed:—

$$C_6H_5.NO_2 + HNO_3 = C_6H_4(NO_2)_2 + H_2O.$$
Di-nitro-benzene.

These nitro bodies are not acids, nor are they ethereal salts of nitrous acid, as nitro-glycerine is of glycerine. They are regarded as formed from nitric acid by the replacement of hydroxyl by benzene radicals.

~Mono-nitro Benzene~ is made by treating benzene with concentrated nitric acid, or a mixture of nitric and sulphuric acids. The latter, as in the case of the nitration of glycerine, takes no part in the reaction, but only prevents the dilution of the nitric acid by the water formed in the reaction. Small quantities may be made thus:— Take 150 c.c. of H_2SO_4 and 75 c.c. HNO_3, or 1 part nitric to 2 parts sulphuric acid, and put in a beaker standing in cold water; then add 15 to 20 c.c. of benzene, drop by drop, waiting between each addition for the completion of the reaction, and shake well during the operation. When finished, pour contents of beaker into about a litre of cold water; the nitro-benzol will sink to the bottom. Decant the water, and wash the nitro-benzol two or three times in a separating funnel with water. Finally, dry the product by adding a little granulated calcium chloride, and allowing to stand for some little time, it may then be distilled. Nitro-benzene is a heavy oily liquid which boils at 205° C., has a specific gravity of 1.2, and an odour like that of oil of bitter almonds. In the arts it is chiefly used in the preparation of aniline.

~Di-nitro Benzene~ is a product of the further action of nitric acid on benzene or nitro-benzene. It crystallises in long fine needles or thin rhombic plates, and melts at 89.9° C. It can be made thus:—The acid mixture used consists of equal parts of nitric and sulphuric acids, say 50 c.c. of each, and without cooling add very slowly 10 c.c. of benzene from a pipette. After the action is over, boil the mixture for a short time, then pour into about half a litre of water, filter off the crystals thus produced, press between layers of filter paper, and crystallise from alcohol. Di-nitro-benzene, or meta-di-nitro-benzene, as it is sometimes called, enters into the composition of several explosives, such as tonite No. 3, roburite, securite, bellite.

Nitro-benzene is manufactured upon the large scale as follows:— Along a bench a row of glass flasks, containing 1 gallon each (1 to 2 lbs. benzene), are placed, and the acids added in small portions at a time, the workmen commencing with the first, and adding a small quantity to each in turn, until the nitration was complete. This process was a dangerous one, and is now obsolete. The first nitro-benzene made commercially in England, by Messrs Simpson, Maule, and Nicholson, of Kennington, in 1856, was by this process. Now, however, vertical iron cylinders, made of cast-iron, are used for the nitrating operation. They are about 4 feet in diameter and 4 feet deep, and a series are generally arranged in a row, at a convenient height from the ground, beneath a line of shafting. Each cylinder is covered with a cast-iron lid having a raised rim all round. A central orifice gives passage to a vertical shaft, and two or more other conveniently arranged openings allow the benzene and the mixed acids to flow in. Each of these openings is surrounded with a deep rim, so that the whole top of the cylinder can be flooded with water some inches in depth, without any of it running into the interior of the nitrator. The lid overhangs the cylinder somewhat, and in the outer rim a number of shot- holes or tubes allow the water to flow down all over the outside of the cylinder into a shallow cast-iron dish, in which it stands. By means of a good supply of cold water, the top, sides, and bottom of the whole apparatus is thus cooled and continually flooded. The agitator consists of cast-iron arms keyed to a vertical shaft, with fixed arms or dash-plates secured to the sides of the cylinder. The shaft has a mitre wheel keyed on the top, which works into a corresponding wheel on the horizontal shafting running along the top of the converters. This latter is secured to a clutch; and there is a feather on the shaft, so that any one of the converters can if necessary be put either in or out of gear. This arrangement is necessary, as riggers or belts of leather, cotton, or indiarubber will not stand the atmosphere of the nitro-benzole house. Above and close to each nitrator stands its acid store tank, of iron or stoneware.

The building in which the nitration is carried out should consist of one story, have a light roof, walls of hard brick, and a concrete floor of 9 to 12 inches thick, and covered with pitch, to protect its surface from the action of the acids. The floor should be inclined to a drain, to save any nitro-benzol spilt. Fire hydrants should be placed at convenient places, and it should be possible to at once fill the building with steam. A 2-inch pipe, with a cock outside the building, is advisable. The building should also be as far as possible isolated.

The acids are mixed beforehand, and allowed to cool before use. The nitric acid used has a specific gravity of 1.388, and should be as free as possible from the lower oxides of nitrogen. The sulphuric acid has a specific gravity of 1.845, and contains from 95 to 96 per cent. of mono- hydrate. A good mixture is 100 parts of nitric to 140 parts of sulphuric acid, and 78 parts of benzene; or 128 parts HNO_3, 179 of H_2SO_4, and 100 of benzene (C_6H_6). The benzene having been introduced into the cylinder, the water is turned on and the apparatus cooled, the agitators are set running, and the acid cock turned on so as to allow it to flow in a very thin stream into the nitrator.

Should it be necessary to check the machinery even for a moment, the stream of acid must be stopped and the agitation continued for some time, as the action proceeds with such vigour that if the benzene being nitrated comes to rest and acid continues to flow, local heating occurs, and the mixture may inflame. Accidents from this cause have been not infrequent. The operation requires between eight to ten hours, agitation and cooling being kept up all the time. When all the acid is added the water is shut off, and the temperature allowed to rise a little, to about 100° C. When it ceases to rise the agitators are thrown out of gear, and the mixture allowed some hours to cool and settle. The acid is then drawn off, and the nitro-benzene is well washed with water, and sometimes distilled with wet steam, to recover a little unconverted benzene and a trace of paraffin (about .5 per cent. together). At many English works, 100 to 200 gallons, or 800 to 1,760 lbs., are nitrated at a time, and toluene is often used instead of benzene, especially if the nitro-benzene is for use as essence of myrbane. The waste acids, specific gravity 1.6 to 1.7, contain a little nitro-benzene in solution and some oxalic acid. They are concentrated in cast-iron pots and used over again.

~Di-nitro Benzene~ is obtained by treating a charge of the hydrocarbon benzene with double the quantity of mixed acids in two operations, or rather in two stages, the second lot of acid being run in directly after the first. The cooling water is then shut off, and the temperature allowed to rise rapidly, or nitro-benzene already manufactured is taken and again nitrated with acids. A large quantity of acid fumes come off, and some of the nitro- and di-nitro-benzol produced comes off at the high temperature which is attained, and a good condensing apparatus of stoneware must be used to prevent loss. The product is separated from the acids, washed with cold water and then with hot. It is slightly soluble in water, so that the washing waters must be kept and used over again. Finally it is allowed to settle, and run while still warm into iron trays, in which it solidifies in masses 2 or 4 inches thick. It should not contain any nitro-benzol, nor soil a piece of paper when laid on it, should be well crystallised, fairly hard, and almost odourless. The chief product is meta-di-nitro-benzene, melting point 89.8, but ortho-di-nitro-benzene, melting point 118°, and para-di-nitro, melting point 172°, are also produced. The melting point of the commercial product is between 85° to 87° C.

Di-nitro-toluene is made in a similar manner. The tri-nitro-benzene can only be made by using a very large excess of the mixed acids. Nitro- benzene, when reduced with iron, zinc, or tin, and hydrochloric acids, forms aniline.

~Roburite.~ — This explosive is the invention of a German chemist, Dr Carl Roth (English patent 267A, 1887), and is now manufactured in England, at Gathurst, near Wigan. It consists of two component parts, non-explosive in themselves (Sprengel's principle), but which, when mixed, form a powerful explosive. The two substances are ammonium nitrate and chlorinated di-nitro-benzol. Nitro-naphthalene is also used. Nitrate of soda and sulphate of ammonium are allowed to be mixed with it. The advantages claimed for the introduction of chlorine into the nitro compound are that chlorine exerts a loosening effect upon the NO_2 groups, and enables the compound to burn more rapidly than when the nitro groups alone are present.

The formula of chloro-di-nitro-benzol is $C_6H_3Cl(NO_2)_2$. The theoretical percentage of nitrogen, therefore, is 13.82, and of chlorine 17.53. Dr Roth states that, from experiments he has made, the dynamic effect is considerably increased by the introduction of chlorine into the nitro compound. Roburite burns quickly, and is not sensitive to shock; it must be used dry; it cannot be made to explode by concussion, pressure, friction, fire, or lightning; it does not freeze; it does not give off deleterious fumes, and it is to all intents and purposes flameless; and when properly tamped and fired by electricity, can be safely used in fiery mines, neither fine dust nor gases being ignited by it. The action is rending and not pulverising. Compared to gunpowder, it is more powerful in a ratio ranging from 2-1/2 to 4 to 1, according to the substance acted upon. It is largely used in blasting, pit sinking, quarrying, &c., but especially in coal mining. According to Dr Roth, the following is the equation of its decomposition: —

$$C_6H_3Cl(NO_2)_2 + 9HN_4NO_3 = 6CO_2 + 20N + HCl.$$

In appearance roburite is a brownish yellow powder, with the characteristic smell of nitro-benzol. Its specific gravity is 1.40. The Company's statement that the fumes of roburite were harmless having been questioned by the miners of the Garswood Coal and Iron Works Colliery, a scientific committee was appointed by the management and the men jointly for the purpose of settling the question. The members of this committee were Dr N. Hannah, Dr D.J. Mouncey, and Professor H.B. Dixon, F.R.S., of Owens College. After a protracted investigation, a long and technical report was issued, completely vindicating the innocuousness of roburite when properly used. In the words of *The Iron and Coal Trades' Review* (May 24, 1889), "The verdict, though not on every point in favour of the use in all circumstances of roburite in coal mines, is yet of so pronounced a character in its favour as an explosive that it is impossible to resist the conclusion that the claims put forward on its behalf rest on solid grounds."

Roburite was also one of the explosives investigated by the committee appointed in September 1889 by the Durham Coalowners' and Miners' Associations, for the purpose of determining whether the fumes produced by certain explosives are injurious to health. Both owners and workmen were represented on the committee, which elected Mr T. Bell, H.M. Inspector of Mines, as its chairman, with Professor P.P. Bedson and Drs Drummond and Hume as professional advisers. The problem considered was whether the fumes produced by the combustion of certain explosives, one of which was roburite, were injurious to health. The trial comprised the chemical analysis of the air at the "intake," and of the vitiated air during the firing of the shots at the "return," and also of the smoky air in the vicinity of the shot-holes. Five pounds and a half of roburite were used in twenty-three shots. It had been asserted that the fumes from this explosive contained carbon-monoxide, CO, but no trace of this gas could be discovered after the explosion. On another occasion, however, when 4.7 lbs. of roburite were exploded in twenty-three shots, the air at the "return" showed traces of CO gas to the extent of .042 to .019 per cent. The medical report which Drs Hume and Drummond presented to the committee shows that they investigated every case of suspected illness produced by exposure to fumes, and they could find no evidence of acute illness being caused. They say, "No case of acute illness has, throughout the inquiry, been brought to our knowledge, and we are led to the conclusion that such cases have not occurred."

~Manufacture.~ — As now made, roburite is a mixture of ammonium nitrate and chlorinated di-nitro-benzol. The nitrate of ammonia is first dried and ground, and then heated in a closed steam-jacketed vessel to a temperature of 80° C., and the melted organic compound is added, and the whole stirred until an intimate mixture is obtained. On cooling, the yellow powder is ready for use, and is stored in straight canisters or made up into cartridges. Owing to the deliquescent nature of the nitrate of ammonia, the finished explosive must be kept out of contact with the air, and for this reason the cartridges are waterproofed by dipping them in melted wax. Roburite is made in Germany, at Witten, Westphalia; and also at the English Company's extensive works at Gathurst, near Wigan, which have been at work now for some eighteen years, having started in 1888. These works are of considerable extent, covering 30 acres of ground, and are equal to an output of 10 tons a day. A canal runs through the centre, separating the chemical from the explosive portions of the works, and the Lancashire and Yorkshire Railway runs up to the doors. Besides sending large quantities of roburite itself abroad, the Company also export to the various colonies the two components, as manufactured in the chemical works, and which separately are quite non-explosive, and which, having arrived at their destination, can be easily mixed in the proper proportions.

Among the special advantages claimed for roburite are: — First, that it is impossible to explode a cartridge by percussion, fire, or electric sparks. If a cartridge or layer be struck with a heavy hammer, the portion struck is decomposed, owing to the large amount of heat developed by the blow. The remaining explosive is not in the least affected, and no detonation whatever takes place. If roburite be mixed with gunpowder, and the gunpowder fired, the explosion simply scatters the roburite without affecting it in the least. In fact, the only way to explode roburite is to detonate it by means of a cap of fulminate, containing at least 1 gramme of fulminate of mercury. Secondly, its great safety for use in coal mines.

Roburite has the great advantage of exploding by detonation at a very low temperature, indeed so low that a very slight amount of tamping is required when fired in the most explosive mixture of air and coal gas possible, and not at all in a mixture of air and coal dust—a condition in which the use of gunpowder is highly dangerous.

Mr W.J. Orsman, F.I.C., in a paper read at the University College, Nottingham, in 1893, gives the temperature of detonation of roburite as below 2,100° C., and of ammonium nitrate as 1,130° C., whereas that of blasting gelatine is as much as 3,220° C. With regard to the composition of the fumes formed by the explosion of roburite, Mr Orsman says: "With certain safety explosives—roburite, for instance—an excess of the oxidising material is added, namely, nitrate of ammonia; but in this case the excess of oxygen here causes a diminution of temperature, as the nitrate of ammonia on being decomposed absorbs heat. This excess of oxygen effectually prevents the formation of carbon monoxide (CO) and the oxides of nitrogen."

The following table (A), also from Mr Orsman's paper, gives the composition of five prominent explosives, and shows the composition of the gases formed on explosion. The gases were collected after detonating 10 grms. of each in a closed strong steel cylinder, having an internal diameter of 5 inches.

With respect to the influence of ammonium nitrate in lowering the temperature of explosion of the various substances to which it is added, it was found by a French Commission that, when dry and finely powdered, ammonium nitrate succeeds in depreciating the heat of decomposition without reducing the power of the explosive below a useful limit. The following table (B) shows the composition of the explosives examined, and the temperatures which accompanied their explosion.

A

Explosive.	Volume of Gas	Composition of Gases.			

	formed. c.c.	CO_2 Per cent.	CO Per cent.	CH_4 &H Per cent.	N. Per cent.
Gunpowder— Nitre 75 parts Sulphur 10 " Charcoal 15 "	2,214	51.3	3.5	3.5	41.7
Gelignite— Nitro-glycerine 56.5 parts Nitro-cotton 3.5 " Wood-meal 8.0 " KNO_3 32.0 "	4,980	25	7	...	67
Tonite— Nitro-Cotton Barium nitrate	3,750	30	8	...	62
Roburite— Ammonium nitrate, 86 parts Di-nitro-chloro-benzol 14 "	4,780	32	...	...	68
Carbonite Nitro-glycerine 25 parts Wood-meal 40 " Potas. nitrate 34 "	2,100	19	15	26	...

B

Explosive	$NH_4.NO_3$ added	Original Temperature	Percentage	Final Temperature	Explosive Co-efficient	Co-efficient
Nitro-glycerine				3,200	...	...
Blasting gelatine (8 per cent. gun-cotton)				3,090	88	1,493
Dynamite (25 per cent. silica)				2,940	80	1,468

| | | | | | | Gun-cotton | 1 | 2,650 | ... | ... | | | 2,060 | 90.5 |
| 1,450 | | | | | | | Ammonium nitrate | 1,130 | ... | ... | | | | |

~Bellite~ is the patent of Mr Carl Lamm, Managing Director of the Rötebro Explosive Company, of Stockholm, and is licensed for manufacture in England. It consists of a mixture of nitrate of ammonia with di- or tri-nitro-benzol, it has a specific gravity of 1.2 to 1.4 in its granulated state, and 1 litre weighs 800 to 875 grms. Heated in an open vessel, bellite loses its consistency at 90° C., but does not commence to separate before a temperature of 200° C. is reached, when it evaporates without exploding. If heated suddenly, it burns with a sooty flame, somewhat like tar, but if the source of heat is removed, it will cease burning, and assume a caramel-like structure. It absorbs very little moisture from the air after it has been pressed, and if the operation has been performed while the explosive is hot, the subsequent increase of weight is only 2 per cent. When subjected to the most powerful blow with a steel hammer upon an iron plate, it neither explodes nor ignites. A rifle bullet fired into it at 50 yards' distance will not explode it. Granulated bellite explodes fully by the aid of fulminating mercury. Fifteen grms. of bellite fired by means of fulminate, projected a shot from an ordinary mortar, weighing 90 lbs., a distance of 75 yards, 15 grms. of gunpowder, under the same conditions, throwing it only 12 yards. A weight of 7-1/2 lbs. falling 145 centimetres failed to explode 1 grm. of bellite.

Various experiments and trials have been made with this explosive by Professor P.T. Cleve, M.P.F. Chalon, C.N. Hake, and by a committee of officers of the Swedish Royal Artillery. It is claimed that it is a very powerful and extremely safe explosive; that it cannot be made to explode by friction, shock, or pressure, nor by electricity, fire, lightning, &c., and that it is specially adapted for use in coal mines, &c.; that it can only be exploded by means of a fulminate detonator, and is perfectly safe to handle and manufacture; that it does not freeze, can be used as a filling for shells, and lastly, can be cheaply manufactured.

~Securite~ consists of 26 parts of meta-di-nitro-benzol and 74 parts of ammonium nitrate. It is a yellow powder, with an odour of nitro-benzol. It was licensed in 1886. It sometimes contains tri-nitro-benzol, and tri-nitro-naphthalene. The equation of its combustion is given as

$$C_6H_4(NO_2)_2 + 10(NH_4NO_3) = 6CO_2 + 22H_2O + 11N_2$$

and, like bellite and roburite, it is claimed to be perfectly safe to use in the presence of fire damp and coal dust.[A] The variety known as Flameless Securite consists of a mixture of nitrate and oxalate of ammonia and di-nitro-benzol.

[Footnote A: See paper by S.B. Coxon, *North of Eng. Inst. Mining and Mech. Eng.*, 11, 2, 87.]

~Kinetite.~—A few years ago an explosive called "Kinetite"[A] was introduced, but is not manufactured in England. It was the patent of Messrs Petry and Fallenstein, and consisted of nitro-benzol, thickened or gelatinised by the addition of some collodion-cotton, incorporated with finely ground chlorate of potash and precipitated sulphide of antimony. An analysis gave the following percentages:—

Nitro-benzol, 19.4 per cent.
Chlorate of potash, 76.9 per cent.
Sulphide of antimony nitro-cotton, 3.7 per cent.

[Footnote A: V. Watson Smith, *Jour. Soc. Chem. Ind.*, January 1887.]

It requires a very high temperature to ignite it, and cannot, under ordinary circumstances, when unconfined, be exploded by the application of heat. It is little affected by immersion in water, unless prolonged, when the chlorate dissolves out, leaving a practical inexplosive residue.[A] It was found to be very sensitive to combined friction and percussion, and to be readily ignited by a glancing blow of wood upon wood. It was also deficient in chemical stability, and has been known to ignite spontaneously both in the laboratory and in a magazine. It is an orange- coloured plastic mass, and smells of nitro-benzol.

[Footnote A: Col. Cundill, R.A., "Dict. of Explosives," says: "If, however, it be exposed to moist and dry air alternately, the chlorate crystallises out on the surfaces, and renders the explosive very sensitive."]

~Tonite No. 3~ contains 10 to 14 per cent. of nitro-benzol (see Tonite). Trench's Flameless Explosive contains 10 per cent. of di-nitro-benzol, together with 85 per cent. of nitrate of ammonia, and 5 per cent. of a mixture of alum, and the chlorides of sodium and ammonia.

~Tri-nitro-Toluene.~ — Toluene, C_7H_8, now chiefly obtained from coal- tar, was formerly obtained by the dry distillation of tolu-balsam. It may be regarded as methyl-benzene, or benzene in which one hydrogen is replaced by methyl (CH_3), thus ($C_6H_5CH_3$), or as phenyl- methane, or methane in which one hydrogen atom is replaced by the radical phenyl (C_6H_5), thus ($CH_3C_6H_5$). Toluene is a colourless liquid, boiling at 110° C., has a specific gravity of .8824 at 0° C., and an aromatic odour. Tri-nitro-toluene is formed by the action of nitric acid on toluene. According to Häussermann, it is more advantageous to start with the ortho-para-di-nitro-toluene, which is prepared by allowing a mixture of 75 parts of 91 to 92 per cent. nitric acid and 150 parts of 95 to 96 per cent. sulphuric acid to run in a thin stream into 100 parts of para-nitro-toluene, while the latter is kept at a temperature between 60° to 65° C., and continually stirred. When the acid has all been run in, this mixture is heated for half an hour to 80° C., and allowed to stand till cold. The excess of nitric acid is then removed. The residue after this treatment is a homogeneous crystalline mass of ortho-para-di-nitro- toluene, of which the solidifying point is 69.5° C. To convert this mass into tri-nitro derivative, it is dissolved by gently heating it with four times its weight of sulphuric acid (95 to 96 per cent.), and it is then mixed with 1-1/2 times its weight of nitric acid (90 to 92 per cent.), the mixture being kept cool. Afterwards it is digested at 90° to 95° C., with occasional stirring, until the evolution of gas ceases. This takes place in about four or five hours.

The operation is now stopped, the product allowed to cool, and the excess of nitric acid separated from it. The residue is then washed with hot water and very dilute soda solution, and allowed to solidify without purification. The solidifying point is 70° C., and the mass is then white, with a radiating crystalline structure. Bright sparkling crystals, melting at 81.5° C. may, however, be obtained by recrystallisation from hot alcohol. The yield is from 100 parts di-nitro-toluene, 150 parts of the tri-nitro derivative. Häussermann states also that 1:2:4:6 tri-nitro- toluene can be obtained from ordinary commercial di-nitro-toluene melting at 60° to 64° C.; but when this is used, greater precautions must be exercised, for the reactions are more violent. Moreover, 10 per cent. more nitric acid is required, and the yield is 10 per cent. less. He also draws attention to the slight solubility of tri-nitro-toluene in hot water, and to the fact that it is decomposed by dilute alkalies and alkaline carbonates—facts which must be borne in mind in washing the substance. This material is neither difficult nor dangerous to make. It behaves as a very stable substance when exposed to the air under varying conditions of temperature (-10° to +50° C.) for several months. It cannot be exploded by flame, nor by heating it in an open vessel. It is only slightly decomposed by strong percussion on an anvil. A fulminate detonator produces the best explosive effect with tri-nitro-toluene. It can be used in conjunction with ammonium nitrate, but such admixture weakens the explosive power; but even then it is stated to be stronger than an equivalent mixture of di-nitro-benzene and ammonium nitrate. Mowbray patented a mixture of 3 parts nitro-toluol to 7 of nitro-glycerine, also in the proportions of 1 to 3, which he states to be a very safe explosive.

~Faversham Powder.~—One of the explosives on the permitted list (coal mines) is extensively used, and is manufactured by the Cotton Powder Co. Ltd. at Faversham. It is composed of tri-nitro-toluol 11 parts, ammonium nitrate 93 parts, and moisture 1 part. This explosive must be used only when contained in a case of an alloy of lead, tin, zinc, and antimony thoroughly waterproof; it must be used only with a detonator or electric detonator of not less strength than that known as No. 6.

~Nitro-Naphthalene.~—Nitro-naphthalene is formed by the action of nitric acid on naphthalene ($C_{10}H_{8}$). Its formula is

$C_{10}H_{7}NO_{2}$, and it forms yellow needles, melting at 61° C.; and of di-nitro-naphthalene ($C_{10}H_{6}(NO_{2})_{2}$), melting point 216° C. There are also tri-nitro and tetra-nitro and [alpha] and [beta] derivatives of nitro-naphthalene. It is the di-nitro-naphthalene that is chiefly used in explosives. It is contained in roburite, securite, romit, Volney's powder, &c. Fehven has patented an explosive consisting of 10 parts of nitro-naphthalene mixed with the crude ingredients of gunpowder as follows:—Nitro-naphthalene, 10 parts; saltpetre, 75 parts; charcoal, 12.5 parts; and sulphur, 12.5 parts. He states that he obtains a mono-nitro-naphthalene, containing a small proportion of di-nitro-naphthalene, by digesting 1 part of naphthalene, with or without heat, in 4 parts of nitric acid (specific gravity 1.40) for five days.

Quite lately a patent has been taken out for a mixture of nitro-naphthalene or di-nitro-benzene with ammonium nitrate, and consists in using a solvent for one or other or both of the ingredients, effected in a wet state, and then evaporating off the solvent, care being taken not to melt the hydrocarbon. In this way a more intimate mixture is ensured between the particles of the components, and the explosive thus prepared can be fired by a small detonator, viz., by 0.54 grms. of fulminate. Favier's explosive also contains mono-nitro-naphthalene (8.5 parts), together with 91.5 parts of nitrate of ammonia. This explosive is made in England by the Miners' Safety Explosive Co. A variety of roburite contains chloro-nitro-naphthalene. Romit consists of 100 parts ammonium nitrate and 7 parts potassium chlorate mixed with a solution of 1 part nitro-naphthalene and 2 parts rectified paraffin oil.

~Ammonite.~—This explosive was originally made at Vilvorde in Belgium, under the title of the Favier Explosive, consisting of a compressed hollow cylinder composed of 91.5 per cent. of nitrate of ammonia, and 8.5 per cent. of mono-nitro-naphthalene filled inside with loose powder of the same composition. The cartridges were wrapped in paper saturated with paraffin-wax, and afterwards dipped in hot paraffin to secure their being water-tight. The Miners' Safety Explosives Co., when making this explosive at their factory at Stanford-le-Hope, Essex, abandoned after a short trial the above composition, and substituted di-nitro-naphthalene 11.5 per cent. for the mono-nitro-naphthalene, and used thin lead envelopes filled with loose powder slightly pressed in, in place of the compressed cylinders containing loose powder. The process of manufacture is shortly as follows:—132-3/4 lbs. of thoroughly dried nitrate of ammonium is placed in a mill pan, heated at the bottom with live steam, and ground for about twenty minutes until it becomes so dry that a slight dust follows the rollers; then 17-1/2 lbs. of thoroughly dry di-nitro-naphthalene is added, and the grinding continued for about ten minutes. Cold water is then circulated through the bottom of the pan until the material appears of a lightish colour and falls to powder. (While the pan is hot the whole mass looks slightly plastic and of a darker colour than when cold.) A slide in the bottom of the pan is then withdrawn, the whole mass working out until the pan is empty; it is now removed to the sifting machine, brushed through a wire sieve of about 12 holes to the inch, and is then ready for filling into cartridges. The hard core is returned from the sifting machine and turned into one of the pans a few minutes before the charge is withdrawn.

The ammonite is filled into the metallic cartridges by means of an archimedian screw working through a brass tube, pushing off the cartridges as the explosive is fed into them against a slight back pressure; a cover is screwed on, and they then go to the dipping room, where they are dipped in hot wax to seal the ends; they are then packed in boxes of 5 lbs. each and are ready for delivery. The di-nitro-naphthalene is made at the factory. Mono-nitro-naphthalene is first made as follows:—12 parts of commercial nitrate of soda are ground to a fine powder, and further ground with the addition of 15 parts of refined naphthalene until thoroughly incorporated; it is then placed in an earthenware pan, and 30 parts of sulphuric acid of 66° B. added, 2 parts at a time, during forty-eight hours (the rate of adding H_2SO_4 depends on the condition of the charge, and keeping it in a fluid state), with frequent agitation, day and night, during the first three or four days, afterwards three or four times a day. In all fourteen days are occupied in the nitration process. It is then strained through an earthenware strainer, washed with warm water, drained, and dried. For the purpose of producing this material in a granulated condition, which is found more convenient for drying, and further nitrification, it is placed in a tub, and live steam passed through, until brought up to the boiling point (the tub should be about half full), cold water is then run in whilst violently agitating the contents until the naphthalene solidifies; it can then be easily drained and dried. For the further treatment to make di-nitro-naphthalene, 18 parts of nitro-naphthalene are placed in an earthenware pan, together with 39 parts of sulphuric acid of 66° B., then 15 parts of nitric acid of 40° B. are added, in small quantities at a time, stirring the mixture continually. This adding of nitric acid is controlled by the fuming, which should be kept down as much as possible. The operation takes ten to twelve days, when 100 times the above quantities, taken in kilogrammes, are taken. At the end of the nitration the di-nitro-naphthalene is removed to earthenware strainers, allowed to drain, washed with hot water and soda until all acid is removed, washed with water and dried. The di-nitro- naphthalene gives some trouble in washing, as some acid is held in the crystals which is liable to make its appearance when crushed. To avoid this it should be ground and washed with carbonate of soda before drying; an excess of carbonate of soda should not, however, be used.

~Electronite.~ — This is a high explosive designed to afford safety in coal getting. This important end has been attained by using such ingredients, and so proportioning them, as will ensure on detonation a degree of heat insufficient under the conditions of a "blown-out" shot, to ignite fire damp or coal dust. It is of the nitrate of ammonium class of permitted explosives. It contains about 75 per cent. of nitrate of ammonium, with the addition of nitrate of barium, wood meal, and starch. The gases resulting from detonation are chiefly water in the gaseous form, nitrogen, and a little carbon dioxide. It is granulated with the object of preventing missfires from ramming, to which nitrate of ammonium explosives are somewhat susceptible. This explosive underwent some exhaustive experiments at the experimental station near Wigan in 1895, when 8 oz. or 12 oz. charges were fired unstemmed into an admixture of coal dust and 10 per cent. of gas, without any ignition taking place. It is manufactured by Messrs Curtis's & Harvey Ltd. at their factory, Tonbridge, Kent.

~Sprengel's Explosives.~ — This is a large class of explosives. The essential principle of them all is the admixture of an oxidising with a combustible agent at the time of, or just before, being required for use, the constituents of the mixture being very often non-explosive bodies. This type of explosive is due to the late Dr Herman Sprengel, F.R.S. Following up the idea that an explosion is a sudden combustion, he submitted a variety of mixtures of oxidising and combustible agents to the violent shock of a detonator of fulminate. These mixtures were made in such proportions that the mutual oxidation or de-oxidation should be theoretically complete. Among them are the following:—

1. One chemical equivalent of nitro-benzene to equivalents of nitric acid.

2. Five equivalents of picric acid to 13 equivalents of nitric acid.

3. Eighty-seven equivalents of nitro-naphthalene to 413 equivalents of nitric acid.

4. Porous cakes, or lumps of chlorate of potash, exploded violently with bisulphide of carbon, nitro-benzol, carbonic acid, sulphur, benzene, and mixtures of these substances.

No. 1 covers the explosive known as *Hellhoffite*, and No. 2 is really oxonite, and No. 4 resembles rack-a-rock, an explosive invented by Mr S.R. Divine, and consisting of a mixture of chlorate of potash and nitro- benzol. Roburite, bellite, and securite should perhaps be regarded as belonging to the Sprengel class of explosives, otherwise this class is not manufactured or used in England. The principal members are known as *Hellhoffite*, consisting of a mixture of nitro-petroleum or nitro-tar oils and nitric acid, or of meta-di-nitro-benzol and nitric acid; *Oxonite*, consisting of picric and nitric acids; and *Panclastite*, a name given to various mixtures, proposed by M. Turpin, such as liquid nitric peroxide, with bisulphide of carbon, benzol, petroleum, ether, or mineral oils.

~Picric Acid, Tri-nitro-Phenol, or Carbazotic Acid.~—Picric acid, or a tri-nitro-phenol $(C_6H_2(NO_2)_3OH)[2:4:6]$, is produced by the action of nitric acid on many organic substances, such as phenol, indigo, wool, aniline, resins, &c. At one time a yellow gum from Botany Bay (*Xanthorrhoea hastilis*) was chiefly used. One part of phenol (carbolic acid), C_6H_5OH, is added to 3 parts of strong fuming nitric acid, slightly warmed, and when the violence of the reaction has subsided, boiled till nitrous fumes are no longer evolved. The resinous mass thus produced is boiled with water, the resulting picric acid is converted into a sodium salt by a solution of sodium carbonate, which throws down sodium picrate in crystals.

Phenol-sulphuric acid is now, however, more generally used, and the apparatus employed for producing it closely resembles that used in making nitro-benzol. It is also made commercially by melting carbolic acid, and mixing it with strong sulphuric acid, then diluting the "sulpho- carbolic"[A] acid with water, and afterwards running it slowly into a stone tank containing nitric acid. This is allowed to cool, where the crude picric acid crystallises out, and the acid liquid (which contains practically no picric acid, but only sulphuric acid, with some nitric acid) being poured down the drains. The crude picric acid is then dissolved in water by the aid of steam, and allowed to cool when most of the picric acid recrystallises. The mother liquor is transferred to a tank and treated with sulphuric acid, when a further crop of picric acid crystals is obtained. The crystals of picric acid are further purified by recrystallisation, drained, and dried at 100° F. on glazed earthenware trays by the aid of steam. It can also be obtained by the action of nitric acid on ortho-nitro-phenol, para-nitro-phenol, and di-nitro-phenol (2:4 and 2:6), but not from meta-nitro-phenol, a fact which indicates its constitution.[B]

[Footnote A: O. and p. phenolsulphonic acids.

$C_3H_4(OH).SO_3H + 3HNO_3 = C_6H_2(NO_2)_3OH + H_2SO_4 + 2H_2O$. (Picric acid).]

[Footnote B: Carey Lea, *Amer. Jour. Sci.*, (ii.), xxxii. 180.]

Picric acid crystallises in yellow shining prisms or laminæ having an intensely bitter taste, and is poisonous. It melts at 122.5° C., sublimes when cautiously heated, dissolves sparingly in cold water, more easily in hot water, still more in alcohol. It stains the skin an intense yellow colour, and is used as a dye for wool and silk. It is a strong acid, forming well crystallised yellow salts, which detonate violently when heated, some of them also by percussion. The potassium salt, $C_6H_2(NO_2)_3OK$, crystallises in long needles very slightly soluble in water. The sodium, ammonium, and barium salts are, however, easily soluble in water. Picric acid, when heated, burns with a luminous and smoky flame, and may be burnt away in large quantity without explosion; but the mere contact of certain metallic oxides, with picric acid, in the presence of heat, develops powerful explosives, which are capable of acting as detonators to an indefinite amount of the acid, wet or dry, which is within reach of their detonative influence. The formula of picric acid is

$$C_6H_2|(NO_2)_3|OH.$$

which shows its formation from phenol ($C_6H_5OH.$), three hydrogen atoms being displaced by the NO_2 group. The equation of its formation from phenol is as follows:—

$$C_6H_5.OH + 3HNO_3 = C_6H_2(NO_2)_3OH + 3H_2O.$$

According to Berthelot, its heat of formation from its elements equals 49.1 calories, and its heat of total combustion by free oxygen is equal to +618.4 cals. It hardly contains more than half the oxygen necessary for its complete combustion.

$$2C_6H_2(NO_2)_3OH + O_{10} = 12CO_2 + 3H_2 + 3N_2.$$

The percentage composition of picric acid is—Nitrogen, 18.34; oxygen, 49.22; hydrogen, 1.00; and carbon, 31.44, equal to 60.26 per cent. of NO_2. The products of decomposition are carbonic acid, carbonic oxide, carbon, hydrogen, and nitrogen, and the heat liberated, according to Berthelot, would be 130.6 cals., or 570 cals. per kilogramme. The reduced volume of the gases would be 190 litres per equivalent, or 829 litres per kilogramme. To obtain a total combustion of picric acid it is necessary to mix with it an oxidising agent, such as a nitrate, chlorate, &c. It has been proposed to mix picric acid (10 parts) with sodium nitrate (10 parts) and potassium bichromate (8.3 parts). These proportions would furnish a third of oxygen in excess of the necessary proportion.

Picric acid was not considered to be an explosive, properly so called, for a long time after its discovery, but the disastrous accident which occurred at Manchester (*vide* Gov. Rep. No. LXXXI., by Colonel (now Sir V.D.) Majendie, C.B.), and some experiments made by Dr Duprè and Colonel Majendie to ascertain the cause of the accident, conclusively proved that this view was wrong. The experiments of Berthelot (*Bull. de la Soc. Chim. de Paris*, xlix., p. 456) on the explosive decomposition of picric acid are also deserving of attention in this connection. If a small quantity of picric acid be heated in a moderate fire, in a crucible, or even in an open test tube, it will melt (at 120° C. commercial acid), then give off vapours which catch fire upon contact with air, and burn with a sooty flame, without exploding. If the burning liquid be poured out upon a cold slab, it will soon go out. A small quantity carefully heated in a tube, closed at one end, can even be completely volatilised without apparent decomposition. It is thus obvious that picric acid is much less explosive than the nitric ethers, such as nitro-glycerol and nitro-cellulose, and very considerably less explosive than the nitrogen compounds and fulminates.

It would, however, be quite erroneous to assume that picric acid cannot explode when simply heated. On the contrary, Berthelot has proved that this is not the case. If a glass tube be heated to redness, and a minute quantity of picric acid crystals be then thrown in, it will explode with a curious characteristic noise. If the quantity be increased so that the temperature of the tube is materially reduced, no explosion will take place at once, but the substance will volatilise and then explode, though with much less violence than before, in the upper part of the tube. Finally, if the amount of picric acid be still further increased under these conditions, it will undergo partial decomposition and volatilise, but will not even deflagrate. Nitrobenzene, di-nitrobenzene, and mono-, di-, and tri-nitronaphthalenes behave similarly.

The manner in which picric acid will decompose is thus dependent upon the initial temperature of the decomposition, and if the surrounding material absorb heat as fast as it is produced by the decomposition, there will be no explosion and no deflagration. If, however, the absorption is not sufficient to prevent deflagration, this may so increase the temperature of the surrounding materials that the deflagration will then end in explosion. Thus, if an explosion were started in an isolated spot, it would extend throughout the mass, and give rise to a general explosion.

In the manufacture of picric acid the first obvious and most necessary precaution is to isolate the substance from other chemicals with which it might accidentally come into contact. If pure materials only are used, the manufacture presents no danger. The finished material, however, must be carefully kept from contact with nitrates, chlorates, or oxides. If only a little bit of lime or plaster become accidentally mixed with it, it may become highly dangerous. A local explosion may occur which might have the effect of causing the explosion of the whole mass. Picric acid can be fired by a detonator, 5-grain fulminate, and M. Turpin patented the use of picric acid, unmixed with any other substance, in 1885. The detonation of a small quantity of dry picric acid is sufficient to detonate a much larger quantity containing as much as 17 per cent. of water.

It is chiefly due to French chemists (and to Dr Sprengel) that picric acid has come to the front as an explosive. Melinite,[A] a substance used by the French Government for filling shells, was due to M. Turpin, and is supposed to be little else than fused picric acid mixed with gun-cotton dissolved in some solvent (acetone or ether-alcohol). Sir F.A. Abel has also proposed to use picric acid, mixed with nitrate of potash (3 parts) and picrate of ammonia (2 parts) as a filling for shells. This substance requires a violent blow and strong confinement to explode it. I am not aware, however, that it has ever been officially adopted in this country. Messrs Désignolles and Brugère have introduced military powders, consisting of mixtures of potassium and ammonium picrates with nitrate of potassium. M. Désignolles introduced three kinds of picrate powders, composed as follows:—

	For Torpedoes and Shells.	For Guns. Ordinary.	For Guns. Heavy.	For Small Arms.
Picrate of Potash	55-50	16.4- 9.6	9	28.6-22.9
Saltpetre	45-50	74.4-79.7	80	65.0-69.4
Charcoal	...	9.2-10.7	11	6.4- 7.7

They were made much like ordinary gunpowder, 6 to 14 per cent. of moisture being added when being milled. The advantages claimed over gunpowder are greater strength, and consequently greater ballistic or disruptive effect, comparative absence of smoke, and freedom from injurious action on the bores of guns, owing to the absence of sulphur. Brugère's powder is composed of ammonium picrate and nitre, the proportions being 54 per cent. picrate of ammonia and 46 per cent. potassic nitrate. It is stable, safe to manufacture and handle, but expensive. It gives good results in the Chassepôt rifle, very little smoke, and its residue is small, and consists of carbonate of potash. It is stated that 2.6 grms. used in a rifle gave an effect equal to 5.5 grms. of ordinary gunpowder.

[Footnote A: The British Lydite and the Japanese Shimose are said to be identical with Melinite.]

Turpin has patented various mixtures of picric acid, with gum-arabic, oils, fats, collodion jelly, &c. When the last-named substance is diluted in the proportion of from 3 to 5 per cent. in a mixture of ether and alcohol, he states that the blocks of picric acid moulded with it will explode in a closed chamber with a priming of from 1 to 3 grammes of fulminate. He also casts picric acid into projectiles, the cast acid having a density of about 1.6. In this state it resists the shock produced by the firing of a cannon, when contained in a projectile, having an initial velocity of 600 metres. It is made in the following way:—The acid is fused in a vessel provided with a false bottom, heated to 130° to 145° C. by a current of steam under pressure, or simply by the circulation under the false bottom of a liquid, such as oil, chloride of zinc, glycerine, &c., heated to the same temperature. The melted picric acid is run into moulds of a form corresponding to that of the blocks required, or it may be run into projectiles, which should be heated to a temperature of about 100° C., in order to prevent too rapid solidification.

When cresylic acid (or cresol, $C_6H_4(CH_3)OH$.) is acted upon by nitric acid it produces a series of nitro compounds very similar to those formed by nitric acids on phenol, such as sodium di-nitro-cresylate, known in the arts as victoria yellow. Naphthol, a phenol-like body obtained from naphthalene, under the same conditions, produces sodium di-nitro- naphthalic acid, $C_{10}H_6(NO_2)_2O$. The explosive known as "roburite" contains chloro-nitro-naphthalene, and romit, a Swedish explosive, nitro-naphthalene.

~Tri-nitro-cresol~, $C_7H_4(NO_2)_3OH$.—A body very similar to tri-nitro-phenol, crystallises in yellow needles, slightly soluble in cold water, rather more so in boiling water, alcohol, and ether. It melts at about 100° C. In France it is known as "Cresilite," and mixed with melinite, is used for charging shells. By neutralising a boiling saturated solution of tri-nitro-cresol with ammonia, a double salt of ammonium and nitro-cresol crystallises out upon cooling, which is similar to ammonium picrate. This salt is known as "Ecrasite," and has been used in Austria for charging shells. It is a bright yellow solid, greasy to the touch, melts at 100° C., is unaffected by moisture, heat, or cold, ignites when brought into contact with an incandescent body or open flame, burning harmlessly away unless strongly confined, and is insensitive to friction or concussion. It is claimed to possess double the strength of dynamite, and requires a special detonator (not less than 2 grms. of fulminate) to provoke its full force. Notwithstanding the excellent properties attributed to this explosive, Lieut. W. Walke ("Lectures on Explosives," p. 181) says, "Several imperfectly explained and unexpected explosions have occurred in loading shells with this substance, and have prevented its general adoption up to the present time."

~The Fulminates.~—The fulminates are salts of fulminic acid, $C_2N_2O_2H_2$. Their constitution is not very well understood. Dr
E. Divers, F.R.S., and Mr Kawakita (*Chem. Soc. Jour.*, 1884, pp. 13-19), give the formulæ of mercury and silver fulminates as

```
  OC = N    AgOC = N
 / | \     | \
Hg | O and | O
 \ | /     | /
  -C = N   AgC = N
```

whereas Dr H.E. Armstrong, F.R.S., would prefer to write the formula of fulminic acid

ON.C.OH. | C(N.OH),

and A.F. Holleman (*Berichte*, v. xxvi., p. 1403), assigns to mercury fulminate the formula

C:N.O
Hg | |
C:N.O,

and R. Schol (*Ber.*, v. xxiii., p. 3505),

C:NO
| | Hg.
C:NO

They are very generally regarded as iso-nitroso compounds.

The principal compound of fulminic acid is the mercury salt commonly known as fulminating mercury. It is prepared by dissolving mercury in nitric acid, and then adding alcohol to the solution, 1 part of mercury and 12 parts of nitric acid of specific gravity 1.36, and 5-1/2 parts of 90 per cent. alcohol being used. As soon as the mixture is in violent reaction, 6 parts more of alcohol are added slowly to moderate the action. At first the mixture blackens from the separation of mercury, but this soon vanishes, and is succeeded by crystalline flocks of mercury fulminate which fall to the bottom of the vessel. During the reaction, large quantities of volatile oxidation products of alcohol, such as aldehyde, ethylic nitrate, &c., are evolved from the boiling liquid, whilst others, such as glycollic acid, remain in solution. The mercury fulminate is then crystallised from hot water. It forms white silky, delicate needles, which are with difficulty soluble in cold water. In the dry state it is extremely explosive, detonating on heating, or by friction or percussion, as also on contact with concentrated sulphuric acid. The reaction that takes place upon its decomposition is as follows:—

$$C_2N_2O_2Hg = Hg + 2CO + N_2 \quad (284)$$

According to this equation 1 grm. of the fulminate should yield 235.8 c.c. (= 66.96 litres for 284 grms.). Berthelot and Vicille have obtained a yield of 234.2 c.c., equal to 66.7 litres for one equivalent 284 grms.

Dry fulminate explodes violently when struck, compressed, or touched with sulphuric acid, or as an incandescent body. If heated slowly, it explodes at 152° C., or if heated rapidly, at 187° C. It is often used mixed with potassium chlorate in detonators. The reaction which takes place in this case is $3C_2N_2O_2Hg + 2KClO_3 = 3Hg + 6CO_2 + 3N_2 + 2KCl$.

On adding copper or zinc to a hot saturated solution of the salt, fulminate of copper or zinc is formed. The copper salt forms highly explosive green crystals. There is also a double fulminate of copper of ammonia, and of copper and potassium. Silver fulminite, $C_2N_2O_2Ag_2$, is prepared in a similar manner to the mercury salt. It separates in fine white needles, which dissolve in 36 parts of boiling water, and are with difficulty soluble in cold water. At above 100° C., or on the weakest blow, it explodes with fearful violence. Even when covered with water it is more sensitive than the mercury salt. It forms a very sensitive double salt with ammonia and several other metals. With hydrogen it forms the acid fulminate of silver. It is used in crackers and bon-bons, and other toy fireworks, in minute quantities. Gay Lussac found it to be composed as follows:—Carbon, 7.92 per cent.; nitrogen, 9.24 per cent.; silver, 72.19 per cent.; oxygen, 10.65 per cent.; and he assigned to it the formula, $C_2N_2Ag_2O_2$. Laurent and Gerhardt give it the formula, $C_2N(NO_2)Ag_2$, and thus suppose it to contain nitryl, NO_2.

On adding potassium chloride to a boiling solution of argentic fulminate, as long as a precipitate of argentic chloride forms, there is obtained on evaporation brilliant white plates, of a very explosive nature, of potassic argentic fulminate, $C(NO_2)KAg.CN$, from whose aqueous solution nitric acid precipitates a white powder of hydric argentic fulminate, $C(NO_2)HAg.CN$. All attempts to prepare fulminic acid, or nitro-aceto-nitrile, $C(NO_2)H_2CN$, from the fulminates have failed. There is a fulminate of gold, which is a violently explosive buff precipitate, formed when ammonia is added to ter-chloride of gold, and fulminate of platinum, a black precipitate formed by the addition of ammonia to a solution of oxide platinum, in dilute sulphuric acid.

Fulminating silver is a compound obtained by the action of ammonia on oxide of silver. It is a very violent explosive. Pure mercury fulminate may be kept an indefinite length of time. Water does not affect it. It explodes at 187° C., and on contact with an ignited body.

It is very sensitive to shock and friction, even that of wood upon wood. It is used for discharging bullets in saloon rifles. Its inflammation is so sudden that it scatters black powder on which it is placed without igniting it, but it is sufficient to place it in an envelope, however weak, for ignition to take place, and the more resisting the envelope the more violent is the shock, a circumstance that plays an important part in caps and detonators. The presence of 30 per cent. of water prevents decomposition, 10 per cent. prevents explosion. This is, however, only true for small quantities, and does not apply to silver fulminate, which explodes under water by friction. Moist fulminates slowly decompose on contact with the oxidisable metals. The (reduced) volume of gases obtained from 1 kilo. is according to Berthelot, 235.6 litres. The equation of its decomposition is $C_2HgN_2O_2 = 2CO + N_2 + Hg$.

Fulminate of mercury is manufactured upon the large scale by two methods. One of these, commonly known as the German method, is conducted as follows:—One part of mercury is dissolved in 12 parts of nitric acid of a specific gravity of 1.375, and to this solution 16.5 parts of absolute alcohol are added by degrees, and heat is then slowly applied to the mixture until the dense fumes first formed have disappeared, and when the action has become more violent some more alcohol is added, equal in volume to that which has already been added. This is added very gradually. The product obtained, which is mercury fulminate, is 112 per cent. of the mercury employed. Another method is to dissolve 10 parts of mercury in 100 parts of nitric acid of a gravity of 1.4, and when the solution has reached a temperature of 54° C, to pour it slowly through a glass funnel into 83 parts of alcohol. When the effervescence ceases, it is filtered through paper filters, washed, and dried over hot water, at a temperature not exceeding 100° C. The fulminate is then carefully packed in paper boxes, or in corked bottles. The product obtained by this process is 130 per cent. of the mercury taken. This process is the safest, and at the same time the cheapest. Fulminate should be kept, if possible, in a damp state. Commercial fulminate is often adulterated with chlorate of potash.

~Detonators~, or caps, are metallic capsules, usually of copper, and resemble very long percussion caps. The explosive is pure fulminate of mercury, or a mixture of that substance with nitrate or chlorate of potash, gun-powder, or sulphur. The following is a common cap mixture:— 100 parts of fulminate of mercury and 50 parts of potassium nitrate, or 100 parts of fulminate and 60 parts of meal powder. Silver fulminate is also sometimes used in caps. There are eight sizes made, which vary in dimensions and in amount of explosive contained. They are further distinguished as singles, doubles, trebles, &c., according to their number. Colonel Cundill, R.A. ("Dict. of Explosives"), gives the following list:—

No. 1 contains 300 grms. of explosive per 1000. " 2 " 400 " " " " " " 3 " 540 " " " " " " 4 " 650 " " " " " " 5 " 800 " " " " " " 6 " 1,000 " " " " " " 7 " 1,500 " " " " " " 8 " 2,000 " " " " "

Trebles are generally used for ordinary dynamite, 5, 6, or 7 for gun-cotton, blasting gelatine, roburite, &c.

In the British service percussion caps, fuses, &c., are formed of 6 parts by weight of fulminate of mercury, 6 of chlorate of potash, and 4 of sulphide of antimony; time fuses of 4 parts of fulminate, 6 of potassium chlorate, 4 of sulphide of antimony, the mixture being damped with a varnish consisting of 645 grains of shellac dissolved in a pint of methylated spirit. Abel's fuse (No. 1) consists of a mixture of sulphide of copper, phosphide of copper, chlorate of potash, and No. 2 of a mixture of gun-cotton and gun-powder. They are detonated by means of a platinum wire heated to redness by means of an electric current. Bain's fuse mixture is a mixture of subphosphide of copper, sulphide of antimony, and chlorate of potash.

In the manufacture of percussion caps and detonators the copper blanks are cut from copper strips and stamped to the required shape. The blanks are then placed in a gun-metal plate, with the concave side uppermost—a tool composed of a plate of gun-metal, in which are inserted a number of copper points, each of the same length, and so spaced apart as to exactly fit each point into a cap when inverted over a plate containing the blanks. The points are dipped into a vessel containing the cap composition, which has been previously moistened with methylated spirit. It is then removed and placed over the blanks, and a slight blow serves to deposit a small portion of the cap mixture into each cap. A similar tool is then dipped into shellac varnish, removed and placed over the caps, when a drop of varnish from each of the copper points falls into the caps, which are then allowed to dry. This is a very safe and efficacious method of working.

At the works of the Cotton-Powder Company Limited, at Faversham, the fulminate is mixed wet with a very finely ground mixture of gun-cotton and chlorate of potash, in about the proportions of 6 parts fulminate, 1 part gun-cotton, and 1 part chlorate. The water in which the fulminate is usually stored is first drained off, and replaced by displacement by methyl-alcohol. While the fulminate is moist with alcohol, the gun-cotton and chlorate mixture is added, and well mixed with it. This mixture is then distributed in the detonators standing in a frame, and each detonator is put separately into a machine for the purpose of pressing the paste into the detonator shell.

At the eleventh annual meeting of the representatives of the Bavarian chemical industries at Regensburg, attention was drawn to the unhealthy nature of the process of charging percussion caps. Numerous miniature explosions occur, and the air becomes laden with mercurial vapours, which exercise a deleterious influence upon the health of the operatives. There is equally just cause for apprehension in respect to the poisonous gases which are evolved during the solution of mercury in nitric acid, and especially during the subsequent treatment with alcohol. Many methods have been proposed for dealing with the waste products arising during the manufacture and manipulation of fulminate of mercury, but according to Kæmmerer, only one of comparatively recent introduction appears to be at all satisfactory. It is based upon the fact that mercuric fulminate, when heated with a large volume of water under high pressure, splits up into metallic mercury and non-explosive mercurial compounds of unknown composition.

In mixing the various ingredients with mercury fulminate to form cap mixtures, they should not be too dry; in fact, they are generally more or less wet, and mixed in small quantities at a time, in a special house, the floors of which are covered with carpet, and the tables with felt. Felt shoes are also worn by the workpeople employed. All the tools and apparatus used must be kept very clean; for granulating, hair sieves are used, and the granulated mixture is afterwards dried on light frames, with canvas trays the bottoms of which are covered with thin paper, and the frames fitted with indiarubber cushions, to reduce any jars they may receive. The windows of the building should be painted white to keep out the rays of the sun.

Mr H. Maxim, of New York, has lately patented a composition for detonators for use with high explosives, which can also be thrown from ordnance in considerable quantities with safety. The composition is prepared as follows:—Nitro-glycerine is thickened with pyroxyline to the consistency of raw rubber. This is done by employing about 75 to 85 per cent. of nitro-glycerine, and 15 to 25 per cent. of pyroxyline, according to the stiffness or elasticity of the compound desired. Some solvent that dissolves the nitro-cotton is also used. The product thus formed is a kind of blasting gelatine, and should be in a pasty condition, in order that it may be mixed with fulminate of mercury. The solvent used is acetone, and the quantity of fulminate is between 75 to 85 per cent. of the entire compound. If desired, the compound can be made less sensitive to shocks by giving it a spongy consistency by agitating it with air while it is still in a syrupy condition. The nitro-glycerine, especially in this latter case, may be omitted. In some cases, when it is desirable to add a deterring medium, nitro-benzene or some suitable gum is added.

[Illustration: FIG. 34. METHOD OF PREPARING THE CHARGE.]

The method of preparing a blasting charge is as follows:—A piece of Bickford fuse of the required length is cut clean and is inserted into a detonator until it reaches the fulminate. The upper portion of the detonator is then squeezed round the fuse with a pair of nippers. The object of this is not only to secure that the full power of the detonator may be developed, but also to fix the fuse in the cap (Fig. 34). When the detonator, &c., is to be used under water, or in a damp situation, grease or tallow should be placed round the junction of the cap with the fuse, in order to make a water-tight joint. A cartridge is then opened and a hole made in its upper end, and the detonator pushed in nearly up to the top. Gun-cotton or tonite cartridges generally have a hole already made in the end of the charge. Small charges of dry gun-cotton, known as primers, are generally used to explode wet gun-cotton. The detonators (which are often fired by electrical means) are placed inside these primers (Fig. 35).

[Illustration: FIG. 35. PRIMER.]

One of the forms of electric exploders used is shown in Fig. 36. This apparatus is made by Messrs John Davis & Son, and is simply a small hand dynamo, capable of producing a current of electricity of high tension. This firm are also makers of various forms of low tension exploders. A charge having been prepared, as in Fig. 34, insert into the bore-hole one or more cartridges as judged necessary, and squeeze each one down separately with a *wooden* rammer, so as to leave no space round the charge, and above this insert the cartridge containing the fuse and detonator. Now fill up the rest of the bore-hole with sand, gravel, water, or other tamping. With gelatine dynamites a firm tamping may be used, but with ordinary dynamite loose sand is better. The charge is now ready for firing.

[Illustration: FIG. 36.—ELECTRIC EXPLODER.]

CHAPTER VI.

SMOKELESS POWDERS.

Smokeless Powder in General — Cordite — Axite — Ballistite — U.S. Naval
Powder — Schultze's E.G. Powder — Indurite — Vielle Poudre — Rifleite —
Cannonite — Walsrode — Cooppal Powders — Amberite — Troisdorf — Maximite —
Picric Acid Powders, &c., &c.

The progress made in recent years in the manufacture of smokeless powders has been very great. With a few exceptions, nearly all these powders are nitro compounds, and chiefly consist of some form of nitro-cellulose, either in the form of nitro-cotton or nitro-lignine; or else contain, in addition to the above, nitro-glycerine, with very often some such substance as camphor, which is used to reduce the sensitiveness of the explosive. Other nitro bodies that are used, or have been proposed, are nitro-starch, nitro-jute, nitrated paper, nitro-benzene, di-nitro-benzene, mixed with a large number of other chemical substances, such as nitrates, chlorates, &c. And lastly, there are the picrate powders, consisting of picric acid, either alone or mixed with other substances.

The various smokeless powders may be roughly divided into military and sporting powders. But this classification is very rough; because although some of the better known purely military powders are not suited for use in sporting guns, nearly all the manufacturers of sporting powders also manufacture a special variety of their particular explosive, fitted for use in modern rifles or machine guns, and occasionally, it is claimed, for big guns also.

Of the purely military powders, the best known are cordite, ballistite, and the French B.N. powder, the German smokeless (which contains nitro- glycerine and nitro-cotton); and among the general powders, two varieties of which are manufactured either for rifles or sporting guns, Schultze's, the E.C. Powders, Walsrode powder, cannonite, Cooppal powder, amberite, &c., &c.

~Cordite~, the smokeless powder adopted by the British Government, is the patent of the late Sir F.A. Abel and Sir James Dewar, and is somewhat similar to blasting gelatine. It is chiefly manufactured at the Royal Gunpowder Factory at Waltham Abbey, but also at two or three private factories, including those of the National Explosives Company Limited, the New Explosives Company Limited, the Cotton-Powder Company Limited, Messrs Kynock's, &c. As first manufactured it consisted of gun-cotton 37 per cent., nitro-glycerine 58 per cent., and vaseline 5 per cent., but the modified cordite now made consists of 65 per cent. gun-cotton, 30 per cent. of nitro-glycerine, and 5 per cent. of vaseline. The gun-cotton used is composed chiefly of the hexa-nitrate,[A] which is not soluble in nitro- glycerine. It is therefore necessary to use some solvent such as acetone, in order to form the jelly with nitro-glycerine. The process of manufacture of cordite is very similar, as far as the chemical part of the process is concerned, to that of blasting gelatine, with the exception that some solvent for the gun-cotton, other than nitro-glycerine has to be used. Both the nitro-glycerine and the gun-cotton employed must be as dry as possible, and the latter should not contain more than .6 per cent. of mineral matter and not more than 10 per cent. of soluble nitro-cellulose, and a nitrogen content of not less than 12.5 per cent. The dry gun-cotton (about 1 per cent. of moisture) is placed in an incorporating tank, which consists of a brass-lined box, some of the acetone is added, and the machine (Fig. 29), is started; after some time the rest of the acetone is added (20 per cent. in all) and the paste kneaded for three and a half hours. At the end of this time the Vaseline is added, and the kneading continued for a further three and a half hours. The kneading machine (Fig. 29) consists of a trough, composed of two halves of a cylinder, in each of which is a shaft which carries a revolving blade. These blades revolve in opposite directions, and one makes about half the number of revolutions of the other. As the blades very nearly touch the bottom of the trough, any material brought into the machine is divided into two parts, kneaded against the bottom, then pushed along the blade, turned over, and completely mixed. During kneading the acetone gradually penetrates the mixture, and dissolves both the nitro-cellulose and nitro-glycerine, and a uniform dough is obtained which gradually assumes a buff colour. During kneading the mass becomes heated, and therefore cold water is passed through the jacket of the machine to prevent heating the mixture above the normal temperature, and consequent evaporation of the acetone. The top of the machine is closed in with a glass door, in order to prevent as far as possible the evaporation of the solvent. When the various ingredients are formed into a homogeneous mass, the mixture is taken to the press house, where in the form of a plastic mass it is placed in cylindrical moulds. The mould is inserted in a specially designed press, and the cordite paste forced through a die with one or more holes. The paste is pressed out by hydraulic pressure, and the long cord is wound on a metal drum (Fig. 38), or cut into lengths; in either case the cordite is now sent to the drying houses, and dried at a temperature of about 100° F. from three to fourteen days, the time varying with the size. This operation drives off the acetone, and any moisture the cordite may still contain, and its diameter decreases somewhat. In case of the finer cordite, such as the rifle cordite, the next operation is blending. This process consists in mounting ten of the metal drums on a reeling machine similar to those used for yarns, and winding the ten cords on to one drum. This operation is known as "ten-stranding." Furthermore, six "ten-stranded" reels are afterwards wound upon one, and the "sixty-stranded" reel is then ready to be sent away, This is done in order to obtain a uniform blending of the material. With cordite of a larger diameter, the cord is cut into lengths of 12 inches. Every lot of cordite from each manufacturer has a consecutive number, numbers representing the size and one or more initial letters to identify the manufacturer. These regulations do not apply to the Royal Gunpowder Factory, Waltham Abbey. The finished cordite resembles a cord of gutta-percha, and its colour varies from light to dark brown. It should not look black or shrivelled, and should always possess sufficient elasticity to return to its original form after slight bending. Cordite is practically smokeless. On explosion a very thin vapour is produced, which is dissipated rapidly. This smokelessness can be understood from the fact that the products of combustion are nearly all non-condensible gases, and contain no solid products of combustion which would cause smoke. For the same muzzle velocity a smaller charge of cordite than gunpowder is required owing to the greater amount of gas produced. Cordite is very slow in burning compared to gunpowder. For firing blank cartridges cordite chips containing no vaseline is used. The rate at which cordite explodes depends in a measure upon the diameter of the cords, and the pressure developed upon its mechanical state. The sizes of cordite used are given by Colonel Barker, R.A., as follows: —

For the .303 rifle .0375 inch diameter. " 12 Pr. B.L. gun .05 " " " .075 " " 4.7-inch Q.F. gun .100 " " 6-inch Q.F. gun .300 " " heavy guns .40 to .50 "

For rifles the cordite is used in bundles of sixty strands, in field-guns in lengths of 11 to 12 inches, and the thicker cordite is cut up into 14-inch lengths. Colonel Barker says that the effect of heat upon cordite is not greater as regards its shooting qualities than upon black powder, and in speaking of the effect that cordite has upon the guns in which it is used (R.A. Inst.) said that they had at Waltham Abbey a 4.7-inch Q.F. gun that had fired 40 rounds of black powder, and 249 rounds of cordite (58 per cent. nitro-glycerine) and was still in excellent condition, and showed very little sign of action, and also a 12-lb. B.L. gun that had been much used and was in no wise injured.

[Footnote A: The gun-cotton used contains 12 per cent. of soluble gun-cotton, and a nitrogen content of not less than 12.8 to 13.1 per cent.]

[Illustration: Fig. 37 Scale, 1 inch = 1 foot. Single Strand Reel.]

[Illustration: FIG. 38. — "TEN-STRANDING."]

In some experiments made by Captain Sir A. Noble,[A] with the old cordite containing 58 per cent. nitro-glycerine, a charge of 5 lbs. 10 oz. of cordite of 0.2 inch diameter was fired. The mean chamber crusher gauge pressure was 13.3 tons per square inch (maximum 13.6, minimum 12.9), or a mean of 2,027 atmospheres (max. 2,070, min. 1,970). The muzzle velocity was 2,146 foot seconds, and the muzzle energy 1,437 foot tons. A gramme of cordite generated 700 c.c. of permanent gases at 0° C. and 760 mm. pressure. The quantity of heat developed was 1,260 gramme units. In the case of cordite, as also with ballistite, a considerable quantity of aqueous vapour has to be added to the permanent gases formed. A similar trial, in which 12 lbs. of ordinary pebble powder was used, gave a pressure of 15.9 tons per square inch, or a mean of 2,424 atmospheres. It gave a 45-lb. projectile a mean muzzle velocity of 1,839 foot seconds, thus developing a muzzle energy of 1,055 foot tons. A gramme of this powder at 0° C. and 760 mm. generates 280 c.c. of permanent gases, and develops 720 grm. units of heat.

[Footnote A: *Proc. Roy. Soc.*, vol. lii., No. 315.]

In a series of experiments conducted by the War Office Chemical Committee on Explosives in 1891, it was conclusively shown that considerable quantities of cordite may be burnt away without explosion. A number of wooden cases, containing 500 to 600 lbs. each of cordite, were placed upon a large bonfire of wood, and burned for over a quarter of an hour without explosion. At Woolwich in 1892 a brown paper packet containing ten cordite cartridges was fired into with a rifle (.303) loaded with cordite, without the explosion of a single one of them, which shows its insensibility to shock.

With respect to the action of cordite upon guns, Sir A. Noble points out that the erosion caused is of a totally different kind to that of black powder. The surface of the barrel in the case of cordite appears to be washed away smoothly by the gases, and not pitted and eaten into as with black powder. The erosion also extends over a shorter length of surface, and in small arms it is said to be no greater than in the case of black powder. Sir A. Noble says in this connection: "It is almost unnecessary to explain that freedom from rapid erosion is of very high importance in view of the rapid deterioration of the bores of large guns when fired with charges developing very high energies. As might perhaps be anticipated from the higher heat of ballistite, its erosive power is slightly greater than that of cordite, while the erosive power of cordite is again slightly greater than that of brown prismatic. Amide powder, on the other hand, possesses the peculiarity of eroding very much less than any other powder with which I have experimented, its erosive power being only one-fourth of that of the other powders enumerated."

TABLE GIVING SOME OF SIR. A. NOBLE'S EXPERIMENTS.

VELOCITIES OBTAINED.

	In a 40 Cal. Gun.	In a 50 Cal. Gun.	In a 75 Cal. Gun.	In a 100 Cal. Gun.
	Foot Secs.	Foot Secs.	Foot Secs.	Foot Secs.
With cordite 0.4 in. diam.	2,794	2,940	3,166	3,286
" " 0.3 "	2,469	2,619	2,811	2,905
" ballistite 0.3 in. cubes	2,416	2,537	2,713	2,806
" French B.N. for 6-inch guns	2,249	2,360	2,536	2,616
" prismatic amide	2,218	2,342	2,511	2,574

ENERGIES REPRESENTED BY ABOVE VELOCITIES.

	Foot Tons.	Foot Tons.	Foot Tons.	Foot Tons.
Cordite 0.4 inch	5,413	5,994	6,950	7,478
Ballistite 0.3 inch cubes	4,227	4,754	5,479	5,852
French B.N.	4,047	4,463	5,104	5,460
Prismatic amide	3,507	3,862	4.460	4,745

And again, in speaking of his own experiments, he says: "One 4.7-inch gun has fired 1,219 rounds, and another 953, all with full charges of cordite, while a 6-inch gun has fired 588 rounds with full charges, of which 355 were cordite. In the whole of these guns, so far as I can judge, the erosion is certainly not greater than with ordinary powder, and differs from it remarkably in appearance. With ordinary powder a gun, when much eroded, is deeply furrowed (these furrows having a great tendency to develop into cracks), and presents much the appearance in miniature of a very roughly ploughed field. With cordite, on the contrary, the surface appears to be pretty smoothly swept away, while the length of the surface eroded is considerably less."

[Illustration: FIG. 39.—COMPARATIVE PRESSURE CURVES OF CORDITE AND BLACK POWDER. *a*, Charge, 48 lbs. powder; *b*, charge, 13 lbs. 4 oz. cordite; *c*, charge, 13 lbs. 4 oz. powder. Weight of projectile, 100 lbs. in 6-inch gun. M.V. Cordite = 1960 feet seconds.]

The pressures given by cordite compared with those given by black powder in the 6-inch gun will be seen upon reference to Fig. 39, which is taken from Professor V.B. Lewes's paper, read before the Society of Arts; and due to Dr W. Anderson, F.R.S., the Director-General of Ordnance Factories.

It has been found that the erosive effect is in direct proportion to the nitro-glycerine present. The cordite M.D., which contains only 30 per cent. nitro-glycerine, gives only about half the erosive effect of the old service cordite. With regard to the heating effect of cordite and cordite M.D. on a rifle, Mr T.W. Jones made some experiments. He fired fifty rounds of .303 cartridges in fifteen minutes in the service rifle. Cordite raised the temperature of the rifle 270° F., and cordite M.D. 160° F. only.

With regard to the effect of heat upon cordite, there is some difference of opinion. Dr W. Anderson, F.R.S., says that there is no doubt that the effect of heat upon cordite is greater than upon black powder. At a temperature of 110° F. the cordite used in the 4.7-inch gun is considerably affected as regards pressure.

Colonel Barker, R.A., in reply to a question raised by Colonel Trench, R.A. (at the Royal Artillery Institution), concerning the shooting qualities of cordite heated to a temperature of 110° F., said: "Heating cordite and firing it hot undoubtedly does disturb its shooting qualities, but as far as we can see, not much more than gunpowder. I fear that we must always expect abnormal results with heated propellants, either gunpowder or cordite; and when fired hot, the increase in pressure and velocities will depend upon the heat above the normal or average temperature at which firing takes place." Colonel Barker also, in referring to experiments that had been made in foreign climates, said: "Climatic trials have been carried out all over the world, and they have so far proved eminently satisfactory. The Arctic cold of the winter in Canada, with the temperature below zero, and the tropical sun of India, have as yet failed to shake the stability of the composition, or abnormally injure its shooting qualities." Dr Anderson is of opinion that cordite should not be stored in naval magazines near to the boilers. Professor Vivian B. Lewes, in his recent Cantor Lectures before the Society of Arts, suggests that the magazines of warships should be water-jacketed, and maintained at a temperature that does not rise above 100° F.

~Axite.~ — This powder is manufactured by Messrs Kynock Limited, at their works at Witton, Birmingham. The main constituents of cordite are retained although the proportions are altered; ingredients are added which impart properties not possessed by cordite, and the methods of its manufacture have been modified. The form has also been altered. Axite is made in the form of a ribbon, the cross section being similar in shape to a double- headed rail. It is claimed for this powder, that it does not corrode the barrel in the way cordite does, that with equal pressure it gives greatly increased velocity, and therefore flatter trajectory. That the effect of temperature on the pressure and velocity with axite is only half that with cordite. That the maximum flame temperature of axite is considerably less than that of cordite, and the erosive effect is therefore considerably less. That the deposit left in the barrel after firing axite cartridges reduces the friction between the bullet and the barrel. It is therefore practicable to use axite cartridges giving higher velocities than can be employed with cordite, as with such velocities the latter would nickel the barrel by excessive friction. It is also claimed that the accuracy is greatly increased. The following results have been obtained with this same time, and under the same conditions: —

~Axite~ Cartridges with 200-grain bullets.
  Velocity 2,726 F.S.
  Pressure 20.95 tons.

~Axite~ Cartridges with 215-grain bullets.
  Velocity 2,498 F.S.
  Pressure 19.24 tons.

~Axite~ Service Cartridges.
  Velocity 2,179 F.S.
  Pressure 15.76 tons.

~Cordite~ Service Cartridges.
  Velocity 2,010 F.S.

Pressure 15.67 tons.

Five rounds from the Service axite and Service cordite were placed in an oven and heated to a temperature of 110° F. for one hour, and were then fired for pressure. The following results were obtained:—

~Axite.~ ~Cordite.~
Before heating 15.76 tons per sq. in. 15.67 tons per sq. in.
After " 16.73 " " 17.21 " "

———— ————

Increase .97 = 6.1% 1.54 = 9.8%

Average Velocities—
Before heating 2,150 F.S. 2,030 F.S.
After " 2,180 " 2,090 "

———— ————

Increase 30 F.S. = 1-1/2% 60.0 F.S. = 3%

In order to show the accuracy given by axite, seven rounds were fired from a machine rest at a target fixed at 100 yards from a rifle. Six of the seven shots could be covered by a penny piece, the other being just outside. In order to ascertain the relative heat imparted to a rifle by the explosion of axite and cordite, ten rounds each of axite and cordite cartridges were fired from a .303 rifle, at intervals of ten seconds, the temperature of the rifle barrel being taken before and after each series:—

THE RISE IN TEMPERATURE OF THE RIFLE BARREL

With axite was 71° F.
With cordite was 89° F.
Difference in favour of axite 18° F. = 20.2%

The lubricating action of axite is shown by the fact that a series of cordite cartridges fired from a .303 rifle in the ordinary way, followed by a second series, the barrel being lubricated between each shot by firing an axite cartridge alternately with the cordite cartridge. The mean velocity of the first series of cordite cartridges was 1,974 ft. per second; the mean velocity of the second series was 2,071 ft. per second; the increased velocity due to the lubricating effect of axite therefore was 97 ft. per second. This powder, it is evident, has very many very excellent qualities, and considerable advantages over cordite. It is understood that axite is at present under the consideration of the British Government for use as the Service powder.

~Ballistite.~ — Nobel's powder, known as ballistite, originally consisted of a camphorated blasting gelatine, and was made of 10 parts of camphor in 100 parts of nitro-glycerine, to which 200 parts of benzol were then added, and 50 parts of nitro-cotton (soluble) were then steeped in this mixture, which was then heated to evaporate off the benzol, and the resulting compound afterwards passed between steam-heated rollers, and formed into sheets, which were then finally cut up into small squares or other shapes as convenient. The camphor contained in this substance was, however, found to be a disadvantage, and its use discontinued. The composition is now 50 per cent. of soluble nitro-cotton and 50 per cent. of nitro-glycerine. As nitro-glycerine will not dissolve its own weight of nitro-cotton (even the soluble variety), benzol is used as a solvent, but is afterwards removed from the finished product, just as the acetone is removed from cordite. About 1 per cent. of diphenylamine is added for the purpose of increasing its stability.

The colour of ballistite is a darkish brown. It burns in layers when ignited, and emits sparks. The size of the cubes into which it is cut is a 0.2-inch cube. Its density is 1.6. It is also, by means of a special machine, prepared in the form of sheets, after being mixed in a wooden trough fitted with double zinc plates, and subjected to the heating process by means of hot-water pipes. It is passed between hot rollers, and rolled into sheets, which are afterwards put through a cutting machine and granulated. Sir A. Nobel's experiments[A] with this powder gave the following results: — The charge used was 5 lbs. 8 oz., the size of the cubes being 0.2 inch. The mean crusher-gauge pressure was 14.3 tons per square inch (maximum, 2,210; minimum, 2,142), and average pressure 2,180 atmospheres. The muzzle velocity was 2,140 foot seconds, and the muzzle energy 1,429 foot tons. A gramme of ballistite generates 615 c.c. of permanent gases, and gives rise to 1,365 grm. units of heat. Ballistite is manufactured at Ardeer in Scotland, at Chilworth in Surrey, and also in Italy, under the name of Filite, which is in the form of cords instead of cubes. The ballistite made in Germany contained more nitro-cellulose, and the finished powder was coated with graphite. Its use has been discontinued as the Service powder in Germany, but it is still the Service powder in Italy.

[Footnote A: *Proc. Roy. Soc.*, vol. lii., p. 315.]

~U.S. Naval Smokeless Powder.~ — This powder is manufactured at the U.S. Naval Torpedo Station for use in guns of all calibres in the U.S. Navy. It is a nitro-cellulose powder, a mixture of insoluble and soluble nitro- cellulose together with the nitrates of barium and potassium, and a small percentage of calcium carbonate. The proportions in the case of the powder for the 6-inch rapid-fire gun are as follows: — Mixed nitro-cellulose (soluble and insoluble) 80 parts, barium nitrate 15 parts, potassium nitrate 4 parts, and calcium carbonate 1 part. The percentage of nitrogen contained in the insoluble nitro-cellulose must be 13.30±0.15, and in the soluble 11.60±0.15, and the mean nitration strength of the mixture must be 12.75 per cent. of nitrogen. The solvent used in making the powder is a mixture of ether (sp. gr. 0.720) 2 parts, and alcohol (95 per cent. by volume) 1 part. The process of manufacture is briefly as follows:[A] — The soluble and insoluble nitro-cellulose are dried separately at a temperature from 38° to 41° C., until they do not contain more than 0.1 per cent. of moisture. The calcium carbonate is also finely pulverised and dried, and is added to the mixed nitro-celluloses after they have been sifted through a 16-mesh sieve. The nitrates are next weighed out and dissolved in hot water, and to this solution is added the mixture of nitro-celluloses and calcium carbonate with constant stirring until the entire mass becomes a homogeneous paste. This pasty mass is next spread upon trays and re-dried at a temperature between 38° and 48° C., and when thoroughly dry it is transferred to the kneading machine. The ether- alcohol mixture is now added, and the process of kneading begun. It has been found by experiment that the amount of solvent required to secure thorough incorporation is about 500 c.c. to each 500 grms. of dried paste. To prevent loss of solvent due to evaporation, the kneading machine is made vapour light. The mixing or kneading is continued until the resulting greyish-yellow paste is absolutely homogeneous so far as can be detected by the eye, which requires from three to four hours. The paste is next treated in a preliminary press (known as the block press and is actuated by hydraulic power), where it is pressed into a cylindrical mass of uniform density and of such dimensions as to fit it for the final or powder press. The cylindrical masses from the block press are transferred to the final press, whence they are forced out of a die under a pressure of about 500 lbs. per square inch. As it emerges from the final press the powder is in the form of a ribbon or sheet, the width and thickness of which is determined by the dimensions of the powder chamber of the gun in which the powder is to be used. On the inner surface of the die are ribs extending in the direction of the powder as it emerges from the press, the object of these ribs being to score the sheets or ribbons in the direction of their length, so that the powder will yield uniformly to the pressure of the gases generated in the gun during the combustion of the charge. The ribbon or sheet is next cut into pieces of a width and length corresponding to the chamber of the gun for which it is intended, the general rule being that the thickness of the grain (when perfectly dry) shall be fifteen one-thousandths (.015) of the calibre of the gun, and the length equal to the length to fit the powder chamber. Thus, in case of the 6-inch rapid-fire gun the thickness of the grain (or sheet) is 0.09 of an inch and the length 32 inches. The sheets are next thoroughly dried, first between sheets of porous blotting-paper under moderate pressure and at a temperature between 15° C. and 21.5° C. for three days, and then exposed to free circulation of the air at about 21.5° C. for seven days, and finally subjected for a week or longer to a temperature not exceeding 38° C. until they cease to lose weight.

[Footnote A: Lieut. W. Walke, "Lectures on Explosives," p. 330.]

The sheets, when thoroughly dried, are of a uniform yellowish-grey colour, and of the characteristic colloidal consistency; they possess a perfectly smooth surface, and are free from internal blisters or cracks. The temperature of ignition of the finished powder should not be below 172° C., and when subjected to the heat or stability test, it is required to resist exposure to a temperature of 71° C. for thirty minutes without causing discoloration of the test paper.

~W.A. Powder.~ — This powder is made by the American Smokeless Powder Company, and it was proposed for use in the United States Army and Navy. It is made in several grades according to the ballistic conditions required. It consists of insoluble gun-cotton and nitro-glycerine, together with metallic nitrates and an organic substance used as a deterrent or regulator. The details of its manufacture are very similar to those of cordite, with the exception that the nitro-glycerine is dissolved in a portion of the acetone, before it is added to the gun-cotton. The powder is pressed into solid threads, or tubular cords or cylinders, according to the calibre of the gun in which the powder is to be used. As the threads emerge from the press they are received upon a canvas belt, which passes over steam-heated pipes, and deposited in wire baskets. The larger cords or cylinders are cut into the proper lengths and exposed upon trays in the drying-house. The powder for small arms is granulated by cutting the threads into short cylinders, which are subsequently tumbled, dusted, and, if not perfectly dry, again placed upon trays in the drying- house. Before being sent away from the factory, from five to ten lots of 500 lbs. each are mixed in a blending machine, in order to obtain greater uniformity. The colour of the W.A. powder is very light grey, the grains are very uniform in size, dry and hard. The powder for larger guns is of a yellowish colour, almost translucent, and almost as hard as vulcanite. The powder is said to be unaffected by atmospheric or climatic conditions, to be stable, and to have given excellent ballistic results; it is not sensitive to the impact of bullets, and when ignited burns quietly, unless strongly confined.

Turning now to the smokeless powders, in which the chief ingredient is nitro-cellulose in some form (either gun-cotton or nitro-lignine, &c.), one of the first of these was Prentice's gun-cotton, which consisted of nitrated paper 15 parts, mixed with 85 parts of unconverted cellulose. It was rolled into a cylinder. Another was Punshon's gun-cotton powder, which consisted of gun-cotton soaked in a solution of sugar, and then mixed with a nitrate, such as sodium or potassium nitrate. Barium nitrate was afterwards used, and the material was granulated, and consisted of nitrated gun-cotton.

The explosive known as tonite, made at Faversham, was at first intended for use as a gunpowder, but is now only used for blasting.

~The Schultze Powder.~ — One of the earliest of the successful powders introduced into this country was Schultze's powder, the invention of Colonel Schultze, of the Prussian Artillery, and is now manufactured by the Schultze Gunpowder Company Limited, of London. The composition of this powder, as given in the "Dictionary of Explosives" by the late Colonel Cundall, is as follows: —

Soluble nitro-lignine 14.83 per cent.
Insoluble " 23.36 "
Lignine (unconverted) 13.14 "
Nitrates of K and Ba 32.35 "
Paraffin 3.65 "
Matters soluble in alcohol 0.11 "
Moisture 2.56 "

This powder was the first to solve the difficulty of making a smokeless, or nearly smokeless powder which could be used with safety and success in small arms. Previously, gun-cotton had been tried in various forms, and in nearly every instance disaster to the weapon had followed, owing to the difficulty of taming the combustion to a safe degree. But about 1866 Colonel Schultze produced, as the result of experiments, a nitrated wood fibre which gave great promise of being more pliable and more easily regulated in its burning than gun-cotton, and this was at once introduced into England, and the Schultze Gunpowder Company Limited was formed to commence its manufacture, which it did in the year 1868. During the years from its first appearance, Schultze gunpowder has passed through various modifications. It was first made in a small cubical grain formed by cutting the actual fibre of timber transversely, and then breaking this veneer into cubes. Later on improvements were introduced, and the wood fibre so produced was crushed to a fine degree, and then reformed into small irregular grains. Again, an advance was made in the form of the wood fibre used, the fibre being broken down by the action of chemicals under high temperature, and so producing an extremely pure form of woody fibre. The next improvement was to render the grains of the powder practically waterproof and less affected by the atmospheric influences of moisture and dryness, and the last improvement to the process was that of hardening the grains by means of a solvent of nitro-lignine, so as to do away with the dust that was often formed from the rubbing of the grains during transit.

Minor modifications have from time to time also been made, in order to meet the gradual alteration which has taken place during this long period in the manufacture of sporting guns and cartridge cases to be used with this powder, but through all its evolution this Company has adhered to the first idea of using woody fibre in preference to cotton as the basis of their smokeless powder, as experience has confirmed the original opinion that a powder can be thus made less sensitive to occasional differences in loading, and more satisfactory all round than when made from the cotton base. The powder has always been regulated so that bulk for bulk it occupies the same measure as the best black powder, and as regards its weight, just one half of that of black.

The process of manufacture of this powder is briefly as follows: —

Wood of clean growth is treated by the well-known sulphite process for producing pure woody fibre, which is very carefully purified, and this, after drying, is steeped in a mixture of nitric and sulphuric acids, to render it a nitro-compound and the explosive base of the powder. This nitro compound is carefully purified until it stands the very high purity requirements of the Home Office, and is then ground with oxygen-bearing salts, &c., and the whole is formed into little irregular-shaped grains of the desired size, which grains are dried and hardened by steeping in a suitable solvent for the nitro compound, and after finally drying, sifting, &c., the powder is stored in magazines for several months before it is issued. When issued, a very large blend is made of many tons weight, which ensures absolute uniformity in the material.

There is in England a standard load adopted by every one for testing a sporting powder; this charge is 42 grains of powder and 1-1/8 oz. No. 6 shot—this shot fired from a 12-bore gun, patterns being taken at 40 yards, the velocity at any required distance.

The standard muzzle velocity of Schultze gunpowder is 1,220 feet per second.

The mean 40 yards ditto is 875 feet per second.

The mean 20 yards ditto is 1,050 feet per second.

The internal pressure not to exceed 3.5 tons.

This Company also manufactures a new form of powder, known as Imperial Schultze. It is a powder somewhat lighter in gravity; 33 grains occupies the bulk charge, as compared with the 42 grains of the old. It follows in its composition much the lines of the older powder, but it is quite free from smoke, and leaves no residue whatever.

~The E.G. Powder.~ — This is one of the oldest of the nitro powders. It was invented by Reid and Johnson in 1882. It is now manufactured by the E.G. Powder Company Limited, at their factory near Dartford, Kent, and in America by the Anglo-American E.G. Powder Company, at New Jersey. The basis of this powder is a fine form of cellulose, derived from cotton, carefully purified, and freed from all foreign substances, and carefully nitrated. Its manufacture is somewhat as follows: — Pure nitro-cotton, in the form of a fine powder, is rotated in a drum, sprinkled with water, and the drum rotated until the nitro-cotton has taken the form of grains. The grains are then dried and moistened with ether-alcohol, whereby the moisture is gelatinised, and afterwards coloured with aurine, which gives them an orange colour. They are then dried and put through a sieve, in order to separate the grains which may have stuck together during the gelatinising process.

Since its introduction soon after 1881, E.G. powder has undergone considerable modifications, and is now a distinctly different product from a practical point of view. It is now and has been since 1897 what is known as a 33-grain powder, that is to say, the old standard charge of 3 drams by measure for a 12-bore gun weighs 33 grains, as compared with 42 grains for the original E.G. and other nitro powders. This improvement was effected by a reduction of the barium nitrate and the use of nitro- cellulose of a higher degree of nitration, and also more gelatinisation in manufacture. The granules are very hard, and resist moisture to an extent hitherto unattainable by any "bulk" powder.

Irregularities of pressure in loading have also a minimum effect by reason of the hardness of the grains. The colouring matter used is aurine, and the small quantity of nitrate used is the barium salt. The powder is standardised for pressure velocity with Boulengé chronograph,[A] pattern and gravimetric density by elaborate daily tests, and is continually subjected to severe trials for stability under various conditions of storage, the result being that it may be kept for what in practice amount to indefinite periods of time, either in cartridges or in bulk without any alteration being feared. The E.C. powders are used in sporting guns. No. 1 and No. 2 E.C. are not at present manufactured, E.C. No. 3 having taken their place entirely. Since 1890 these powders have been manufactured under the Borland-Johnson patents, these improved powders being for some time known as the J.B. powders. The E.C. No. 1 was superseded by the E.C. No. 2, made under the Borland-Johnson patents, and this in its turn by the E.C. No. 3 (in 1897).

[Footnote A: Invented in 1869 by Major Le Boulengé, Belgian Artillery. It is intended to record the mean velocity between any two points, and from its simplicity and accuracy is largely employed. Other forms have been invented by Capt. Bréger, French Artillerie de la Marine, and Capt. Holden, R.A.]

~Indurite~ is the invention of Professor C.E. Munroe, of the U.S. Naval Torpedo Station. It is made from insoluble nitro-cotton, treated in a particular manner by steam, and mixed with nitro-benzene. The *Dupont* powder is very similar to *Indurite*. M.E. Leonard, of the United States, invented a powder consisting of 75 parts of nitro-glycerine, 25 parts of gun-cotton, 5 parts of lycopodium powder, and 4 parts of urea crystals dissolved in acetone. The French smokeless powder, Vielle poudre (poudre B), used in the Lebel rifle, is a mixture of nitro-cellulose and tannin, mixed with barium and potassium nitrates. It gives a very feeble report, and very little bluish smoke. The Nobel Company is said to be perfecting a smokeless powder in which the chief ingredients are nitro-amido- and tri-nitro-benzene. C.O. Lundholm has patented (U.S. Pat, 701,591, 1901) a smokeless powder containing nitro-glycerine 30, nitro-cellulose 60, diamyl phthalate 10 (or diamyl phthalate 5, and mineral jelly 5). The diamyl phthalate is added, with or without the mineral jelly to nitro-glycerine and nitro-cellulose.

~Walsrode Powder.~ — The smokeless powder known as Walsrode powder consists of absolutely pure gelatinised nitro-cellulose, grained by a chemical not a mechanical process, consequently the grains do not need facing with gelatine to prevent their breaking up, as is the case with many nitro powders. For this same reason, as well as from the method of getting rid of the solvent used, the Walsrode has no tendency whatever to absorb moisture. In fact, it can lie in water for several days, and when taken out and dried again at a moderate temperature will be found as good as before. Nor is it influenced by heat, whether dry or damp, and it can be stored for years without being in the least affected. It is claimed also that it heats the barrels of guns much less than black powder, and does not injure them.

The standard charge is 30 grains, and it is claimed that with this charge Walsrode powder will prove second to none. A large cap is necessary, as the grains of this powder are very hard, and require a large flame to properly ignite them. In loading cartridges for sporting purposes, an extra felt wad is required to compensate for the small space occupied by the charge; but for military use the powder can be left quite loose. The gas pressure of this powder is low (in several military rifles only one- half that of other nitros), and the recoil consequently small; and it is claimed that with the slight increase of the charge (from 29 to 30 grs.) both penetration and initial velocity will be largely increased, whilst the gas pressure and recoil will not be greater.

This powder was used at Bisley, at the National Rifle Association's Meeting, with satisfactory results. It is made by the Walsrode Smokeless and Waterproof Gunpowder Company. The nitro-cotton is gelatinised by means of acetic ether, and the skin produced retards burning. The nitro-cotton is mixed with acetic ether, and when the gelatinisation has taken place, the plastic mass is forced through holes in a metal plate into strips, which are then cut up into pieces the size of grains. The M.H. Walsrode powder is a leaflet powder, light in colour, about 40 grains of which give a muzzle velocity of 1,350 feet and a pressure of 3 tons. It is, like the other Walsrode powders, waterproof and heat-proof.

~Cooppal Powder~ is manufactured by Messrs Cooppal & Co. at their extensive powder works in Belgium. It consists of nitro-jute or nitro- cotton, with or without nitrates, treated with a solvent to form a gelatinised mass. There are a great many varieties of this powder.

One kind is in the form of little squares; another, for use in Hotchkiss guns, is formed into 3-millimetre cubes, and is black. Other varieties are coloured with aniline dyes of different colours.

~Amberite~ is a nitro-cellulose powder of the 42-grain type of sporting gunpowders, and is manufactured by Messrs Curtis's & Harvey Limited, at their Smokeless Powder Factory, Tonbridge, Kent. It consists of a mixture of nitro-cellulose, paraffin, barium, nitrate, and some other ingredients. It is claimed for this powder that it combines hard shooting with safety, great penetration, and moderate strain on the gun. It is hard and tough in grain, and may be loaded like black powder, and subjected to hard friction without breaking into powder, that it is smokeless, and leaves no residue in the gun. The charge for 12 bores is 42 grains by weight, and 1-1/8 oz. or 1-1/16 oz. shot. The powders known as cannonite[A] and ruby powder, also manufactured by Messrs Curtis's & Harvey Limited, are analogous products having the same general characteristics.

[Footnote A: For further details of cannonite, see First Edition, p. 181.]

~Smokeless Diamond~, also manufactured by the above mentioned firm, is a nitro-cellulose powder of the 33-grain type of sporting gunpowders. It was invented by Mr H.M. Chapman. The manufacture of Smokeless Diamond, as carried out at Tonbridge, is shortly as follows:—The gun-cotton, which is the chief ingredient of this powder, is first stoved, then mixed with certain compounds which act as moderators, and after the solvents are added, is worked up into a homogeneous plastic condition. It then undergoes the processes of granulation, sifting, dusting, drying, and glazing. In order to ensure uniformity several batches are blended together, and stored for some time before being issued for use.

It is claimed for this powder that it is quick of ignition, the quickness being probably due to the peculiar structure of the grains which, when looked at under the microscope, have the appearance of coke. The charge for a 12 bore is 33 grains and 1-1/16 oz. shot, which gives a velocity of 1,050 feet per second, and a pressure of 3 tons per square inch.

~Greiner's Powder~ consists of nitro-cellulose, nitro-benzol, graphite, and lampblack.

~B.N. Powder.~ — This powder is of a light grey or drab colour, perfectly opaque, and rough to the touch. It consists of a mixture, nitro-cellulose and the nitrates of barium and potassium. Its composition is as follows: —

Insoluble nitro-cellulose 29.13 parts
Soluble nitro-cellulose 41.31 "
Barium nitrate 19.00 "
Potassium nitrate 7.97 "
Sodium carbonate 2.03 "
Volatile matter 1.43 "

This powder is a modification of the Poudre B., or Vieille's powder invented for use in the Lebel rifle, and which consisted of a mixture of the nitro-celluloses with paraffin.

~Von Foster's Powder~ contains nothing but pure gelatinised nitro- cellulose, together with a small quantity of carbonate of lime.

The German ~Troisdorf Powder~ is a mixture of gelatinised nitro-cellulose, with or without nitrates.

~Maximite~ is the invention of Mr Hudson Maxim, and is a nitro-compound, the base being gun-cotton. The exact composition and method of manufacture are, however, kept secret. It is made by the Columbia Powder Manufacturing Company, of New York, and in two forms — one for use as a smokeless rifle powder, and the other for blasting purposes.

~Wetteren Powder.~ — This powder was manufactured at the Royal Gunpowder Factory at Wetteren, and used in the Belgian service. Originally it was a mixture of nitro-glycerine and nitro-cellulose, with amyl acetate as solvent. Its composition has, however, been altered from time to time. One variety consists chiefly of nitro-cellulose, with amyl acetate as solvent. It is of a dark brown colour, and of the consistency of indiarubber. It is rolled into sheets and finally granulated.

~Henrite~ is a nitro-cellulose powder.

~Normal Powder.~—The Swedish powder known as "Normal" Smokeless Powder, and manufactured by the Swedish Powder Manufacturing Company, of Landskrona, Sweden, and used for some years past in the Swiss Army, is made in four forms. For field guns of 8.4 calibre, it is used in the form of cylindrical grains of a yellow colour, of a diameter of .8 to .9 mm. and density of .790—about 840 grains of it go to one gun. For rifles, it is used in the form of grey squares, density .750, and 1 grm. equals about 1,014 grains. One hundred rounds of this powder, fired in eighteen minutes, raised the temperature of the gun barrel 284° F. A nitro-glycerine powder, fired under the same conditions, gave a temperature of 464° F.

This powder is said to keep well—a sample kept 3-1/2 years gave as good results as when first made—is easy to make, very stable, ignites easily, not very sensitive to shock or friction, is very light, &c. Eight hundred rounds fired from a heavy gun produced no injury to the interior of the weapon. Samples kept for eleven months in the moist atmosphere of a cellar, when fired gave a muzzle velocity of 1,450 ft. secs. and pressure of 1,312 atmospheres, and the moisture was found to have risen from 1.2 to 1.6 per cent. After twenty-three months in the damp it contained 2 per cent. moisture, gave a muzzle velocity of 1,478 ft. sees., and pressure of 1,356 atmospheres. In a 7.5 millimetre rifle, 13.8 grm. bullet, and charge of 2 grms., it gives a muzzle velocity of 2,035 ft. secs. and a pressure of 2,200 atmospheres. In the 8.4 cm. field-gun, with charge of 600 grms., and projectile of 6.7 kilogrammes, muzzle velocity was equal to 1,640 ft. secs. and pressure 1,750. A sample of the powder for use in the .303 M. rifle, lately analysed by the author, gave the following result:—

Gun-cotton 96.21 per cent.
Soluble cotton 1.80 "
Non-nitrated cotton trace.
Resin and other matters 1.99 "

———

100.00

The various forms of powder invented and manufactured by Mr C.F. Hengst are chiefly composed of nitrated straw that has been finely pulped. The straw is treated first with acids and afterwards with alkalies, and the result is a firm fibrous substance which is granulated. It is claimed that this powder is entirely smokeless and flameless, that it does not foul the gun nor heat the barrel, and is at the same time 150 per cent. stronger than black powder.

The German "Troisdorf" powder consists of nitro-cellulose that has been gelatinised together with a nitrate. Kolf's powder is also gelatinised with nitro-cellulose. The powders invented by Mr E.J. Ryves contain nitro- glycerine, nitro-cotton, castor-oil, paper-pulp, and carbonate of magnesia. Maxim powder contains both soluble and insoluble nitro- cellulose, nitro-glycerine, and carbonate of soda. The smokeless powder made by the "Dynamite Actiengesellschaft Nobel" consists of nitro-starch 70 to 99 parts, and of di- or tri-nitro-benzene 1 to 30 parts.

An American wood powder, known as Bracket's Sporting Powder, consists of soluble and insoluble nitro-lignine, mixed with charred lignine, humus, and nitrate of soda. Mr F.H. Snyder, of New York, is the inventor of a shell powder known as the "Snyder Explosive," consisting of 94 per cent. nitro-glycerine, 6 per cent. of soluble nitro-cotton, and camphor, which is said to be safe in use. Experiments were made with it in a 6-inch rifled gun, fired at a target 220 yards away, composed of twelve 1-inch steel plates welded together, and backed with 12-inch and 14-inch oak beams, and weighing 20 tons. The shots entirely destroyed it. The charge of explosive used was 10 lbs. in each shell.

~Comparative Tests of Black and Nitro Powders, from "American Field."~ — The results given in table below were obtained at the German Shooting Association's grounds at Coepenick, Berlin. Penetration was calculated by placing frames, each holding five cards of 1 millimetre in thickness (equals .03937 inch), and 3 inches apart, in a bee-line, at distances of 20 inches. Velocity, pattern, and penetration were taken at 40 yards from the muzzle of a 12-gauge choke-bore double-barrel gun. Gas pressure was taken by a special apparatus. All shells were loaded with 1-1/8 oz. of No. 3 shot, equal to 120 pellets, and the number given below represents the average number in the 30-inch pattern. The number of sheets passed through gives the average penetration. One atmosphere equals pressure equal to 1 kilogramme (2.2 lbs.) on the square centimetre, hence 1,000 atmospheres equal 2,200 lbs. on the square centimetre. The E.C., Schultze, and Walsrode powders were loaded in Elcy's special shells, 2-1/2 inches long. The averages were taken from a large number of shots, and the same series of shots fired under precisely the same conditions.

	Pattern.	Gas Pressure.	Penetration.		Velocity.
		Atmospheres.	Metres.	Sheets.	
Fine-grained black powder, standard charge	514.2	280	78.6 = 66%	19.0	
Coarse-grained black powder, standard charge	473.4	281.4	78.2 = 65%	19.4	
Schultze powder, 42 grains	921.0	290.0	64.2 = 54%	20.2	
Schultze powder, 45 grains	1052.8	305.8	52.2 = 42%	20.6	
E.G. smokeless, 42 grains	920.2	298.4	81.4 = 67%	18.8	
Walsrode, 29 grains	586.4	280.6	83.0 = 69%	19.0	

Barometer, 760 mm. Thermometer, 30° C. Hydrometer = 65. Wind, S.W.

~Picric Powders.~ — The chief of these is *Melinite*, the composition of which is not known with certainty. It is believed to be melted picric acid together with gun-cotton dissolved in acetone or ether-alcohol. Walke gives the following proportions — 30 parts of tri-nitro-cellulose dissolved in 45 parts of ether-alcohol (2 to 1), and 70 parts of fused and pulverised picric acid. The ether-alcohol mixture is allowed to evaporate spontaneously, and the resulting cake granulated. The French claim, however, that the original invention has been so modified and perfected that the melinite of to-day cannot be recognised in the earlier product. Melinite has a yellow colour, is almost without crystalline appearance, and when ignited by a flame or heated wire, it burns with a reddish-yellow flame, giving off copious volumes of black smoke. Melinite as at present used is said to be a perfectly safe explosive, both as regards manufacture, handling, and storage.

Lyddite,[A] the picric acid explosive used in the British service, is supposed to be identical with the original melinite, but its composition has not been made public.

[Footnote A: Schimose, the Japanese powder, is stated to be identical with
Lyddite and Melinite (*Chem. Centr.*, 1906, 1, 1196).]

Picrates are more often used than picric acid itself in powders. One of the best known is *Brugère's Powder*, which is a mixture of 54 parts of picrate of ammonia and 45 parts of saltpetre. It is stable and safe to manufacture. It has been used in the Chassepôt rifle with good results, gives little smoke, and a small residue only of carbonate of potash.

The next in importance is *Designolle's Powder*, made at Bouchon, consisting of picrate of potash, saltpetre, and charcoal. It was made in three varieties, viz., for rifles, big guns, and torpedoes and shells. These powders are made much in the same way as gunpowder. The advantages claimed for them over gunpowder are, greater strength, comparative absence of smoke, and freedom from injurious action on the bores of guns.

Emmensite is the invention of Dr Stephen Emmens, of the United States. The Emmens "crystals" are produced by treating picric acid with fuming nitric acid of specific gravity of 1.52. The acid dissolves with the evolution of red fumes. The liquid, when cooled, deposits crystals, stated to be different to picric acid, and lustrous flakes. These flakes, when heated in water, separate into two new bodies. One of these enters into solution and forms crystals unlike the first, while the other body remains undissolved. The acid crystals are used mixed with a nitrate.

Emmensite has been subjected to experiment by the direction of the U.S. Secretary for War, and found satisfactory. A sample of Emmensite, in the form of a coarse powder, was first tried in a pistol, and proved superior in propelling power to ordinary gunpowder. When tested against explosive gelatine, it did very good work in shattering iron plates. It is claimed for this explosive that it enjoys the distinction of being the only high explosive which may be used both for firearms and blasting. This view is supported by the trials made by the American War Office authorities, and shows Emmensite to be a useful explosive both for blasting and as a smokeless powder. Its explosive power, as tested, is 283 tons per square inch, and its specific gravity is 1.8.

Abel proposed to use picric acid for filling shells. His *Picric Powder* consisted of 3 parts of saltpetre, and 2 of picrate of ammonia. *Victorite* consists of chlorate of potash, picric acid, and olive oil, and with occasionally some charcoal. It has the form of a coarse yellowish grey powder, and leaves an oily stain on paper, and it is very sensitive to friction and percussion. The composition is as follows:— $KClO_3$ = 80 parts; picric acid, 110 parts; saltpetre, 10 parts; charcoal, 5 parts. It is not manufactured in England. *Tschiner's Powder* is very similar to Victorite in composition, but contains resin. A list of the chief picric powders will be found in the late Colonel J.P. Cundill, R.A.'s "Dictionary of Explosives."

CHAPTER VII.

ANALYSIS OF EXPLOSIVES.

Kieselguhr Dynamite—Gelatine Compounds—Tonite—Cordite—Vaseline—
Acetone—Scheme for Analysis of Explosives—Nitro-Cotton—Solubility Test—
Non-Nitrated Cotton—Alkalinity—Ash and Inorganic Matter—Determination
of Nitrogen—Lungé, Champion and Pellet's, Schultze-Tieman, and Kjeldahl's
Methods—Celluloid—Picric Acid and Picrates—Resinous and Tarry Matters—
Sulphuric Acid and Hydrochloric Acid and Oxalic Acid—Nitric Acid—
Inorganic Impurities—General Impurities and Adulterations—Potassium
Picrate, &c.—Picrates of the Alkaloids—Analysis of Glycerine—Residue—
Silver Test—Nitration—Total Acid Equivalent—Neutrality—Free Fatty
Acids—Combined Fatty Acids—Impurities—Oleic Acid—Sodium Chloride—
Determination of Glycerine—Waste Acids—Sodium Nitrate—Mercury
Fulminate—Cap Composition—Table for Correction of Volumes of Gases, for
Temperature and Pressure

~Kieselguhr Dynamite.~—The material generally consists of 75 per cent. of nitro-glycerine and 25 per cent. of the infusorial earth kieselguhr. The analysis is very simple, and may be conducted as follows:—Weigh out about 10 grms. of the substance, and place over calcium chloride in a desiccator for some six to eight days, and then re-weigh. The loss of weight gives the moisture. This will generally be very small, probably never more than 1 per cent., and usually less.

Mr James O. Handy, in order to save time, proposes to dry dynamite in the following manner. He places 1 grm. of the material in a porcelain crucible 1 inch in diameter. The crucible is then supported at the bottom of an extra wide-mouthed bottle of about 600 c.c. capacity. Air, which has been dried by bubbling through strong sulphuric acid, is now drawn over the surface of the sample for three hours by means of an ordinary aspirator. The air should pass approximately at the rate of 10 c.c. per second. The tube by which the dry air enters the bottle extends to within 1 inch of the crucible containing the dynamite. An empty safety bottle is connected with the inlet, and another with the outlet of the wide-mouthed bottle. The first guards against the mechanical carrying over by the air current of sulphuric acid from the acid bottle into the sample, whilst the second prevents spasmodic outbursts of water from the exhaust from reaching the sample. The method also gave satisfactory results with nitro-glycerine. The dry substance may now be wrapped in filter paper, the whole weighed, and the nitro-glycerine extracted in the Soxhlet apparatus with ether. The ether should be distilled over at least twenty-four times.

I have found, however, that much quicker, and quite as accurate, results may be obtained by leaving the dynamite in contact with ether in a small Erlenmeyer flask for twenty-four hours—leaving it overnight is better— and decanting, and again allowing the substance to remain in contact with a little fresh ether for an hour, and finally filtering through a weighed filter, drying at 100° C., and weighing. This gives the weight of the kieselguhr. The nitro-glycerine must be obtained by difference, as it is quite useless to evaporate down the ethereal solution to obtain it, as it is itself volatile to a very considerable extent at the temperature of evaporation of the ether, and the result, therefore, will always be much too low. The dry guhr can, of course, be examined, either qualitatively or quantitatively, for other mineral salts, such as carbonate of soda, &c. An actual analysis of dynamite No. 1 made by the author at Hayle gave— Moisture, 0.92 per cent.; kieselguhr, 26.15 per cent.; and nitro- glycerine, 72.93 per cent., the last being obtained by difference.

~Nitro-Glycerine.~—It is sometimes desired to test an explosive substance for nitro-glycerine. If an oily liquid is oozing from the substance, soak a drop of it in filter paper. If it is nitro-glycerine it will make a greasy spot. If the paper is now placed upon an iron anvil, and struck with an iron hammer, it will explode with a sharp report, if lighted it burns with a yellowish to greenish flame, emitting a crackling sound, and placed upon an iron plate and heated from beneath, it explodes sharply.

If a few drops of nitro-glycerine are placed in a test tube, and shaken up with methyl-alcohol (previously tested with distilled water, to see that it produces no turbidity), and filtered, on the addition of distilled water, the solution will become milky, and the nitro-glycerine will separate out, and finally collect at the bottom of the tube.

If to a solution of a trace of nitro-glycerine in methyl-alcohol, a few drops of a solution, composed of 1 volume of aniline, and 40 volumes sulphuric acid (1.84) be added, a deep purple colour will be produced. This colour changes to green upon the addition of water. If it is necessary to determine the nitro-glycerine quantitatively in an explosive, the scheme on page 213 may be followed. Ether is the best solvent to use. Nitrogen should be determined in the nitrometer.

~Gelatine Compounds.~—The simplest of these compounds is, of course, blasting gelatine, as it consists of nothing but nitro-cotton and nitro- glycerine, the nitro-cellulose being dissolved in the glycerine to form a clear jelly, the usual proportions being about 92 per cent. of nitro- glycerine to 8 per cent. nitro-cotton, but the cotton is found as high as 10 per cent. in some gelatines. Gelatine dynamite and gelignite are blasting gelatines, with varying proportions of wood-pulp and saltpetre (KNO_3) mixed with a thin blasting gelatine. The method of analysis is as follows:—Weigh out 10 grms. of the substance, previously cut up into small pieces with a platinum spatula, and place over calcium chloride in a desiccator for some days. Reweigh. The loss equals moisture. This is generally very small. Or Handy's method may be used. The dried sample is then transferred to a small thistle-headed funnel which has been cut off from its stem, and the opening plugged with a little glass wool, and round the top rim of which a piece of fine platinum wire has been fastened, in order that it may afterwards be easily removed from the Soxhlet tube. The weight of this funnel and the glass wool must be accurately known. It is then transferred to the Soxhlet tube and exhausted with ether, which dissolves out the nitro-glycerine. The weighed residue must afterwards be treated in a flask with ether-alcohol to dissolve out the nitro-cotton.

But the more expeditious method, and one quite as accurate, is to transfer the dried gelatine to a conical Erlenmeyer flask of about 500 c.c. capacity, and add 250 c.c. of a mixture of ether-alcohol (2 ether to 1 alcohol), and allow to stand overnight. Sometimes a further addition of ether-alcohol is necessary. It is always better to add another 300 c.c., and leave for twenty minutes or so after the solution has been filtered off. The undissolved portion, which consists of wood-pulp, potassium nitrate, and other salts, is filtered off through a linen or paper filter, dried and weighed.

~Solution.~ — The ether-alcohol solution contains the nitro-cotton and the nitro-glycerine in solution.[A] To this solution add excess of chloroform (about 100 c.c. will be required), when the nitro-cellulose will be precipitated in a gelatinous form. This should be filtered off through a linen filter, and allowed to drain. It is useless to attempt to use a filter pump, as it generally causes it to set solid. The precipitated cotton should then be redissolved in ether-alcohol, and again precipitated with chloroform (20 c.c. of ether-alcohol should be used). This precaution is absolutely necessary, if the substance has been treated with ether- alcohol at first instead of ether only, otherwise the results will be much too high, owing to the gelatinous precipitate retaining very considerable quantities of nitro-glycerine. The precipitate is then allowed to drain as completely as possible, and finally allowed to dry in the air bath at 40° C., until it is easily detached from the linen filter by the aid of a spatula, and is then transferred to a weighed watch-glass, replaced in the oven, and dried at 40° C. until constant in weight. The weight found, calculated upon the 10 grms. taken, gives the percentage of nitro- cellulose.

[Footnote A: If the substance has been treated with ether alone in the Soxhlet, the nitro-glycerine will of course be dissolved out first, and the ether-alcohol solution will only contain the nitro-cellulose.]

~The Residue~ left after treating the gelatine with ether-alcohol is, in the case of blasting gelatine, very small, and will probably consist of nothing but carbonate of soda. It should be dried at 100° C. and weighed, but in the case of either gelignite or gelatine dynamite this residue should be transferred to a beaker and boiled with distilled water, and the water decanted some eight or ten times, and the residue finally transferred to a tarred filter and washed for some time with hot water. The residue left upon the filter is wood-pulp. This is dried at 100° C. until constant, and weighed. The solution and washings from the wood are evaporated down in a platinum dish, and dried at 100° C. It will consist of the potassium nitrate, and any other mineral salts, such as carbonate of soda, which should always be tested for by adding a few drops of nitric acid and a little water to the residue, and again evaporating to dryness and re-weighing. From the difference in weight the soda can be calculated, sodium nitrate having been formed. Thus—

$Na_2CO_3 + 2HNO_3 = 2NaNO_3 + CO_2 + H_2O$.

Mol. wt. = 106 = 170

(170 - 106 = 64) and $x = (106 \times d)/64$

where x equals grms. of sodium carbonate in residue, and d equals the difference in weight of residue, before and after treatment with nitric acid.

The nitro-glycerine is best found by difference, but if desired the solutions from the precipitation of the nitro-cellulose may be evaporated down upon the water bath at 30° to 40° C., and finally dried over $CaCl_2$ until no smell of ether or chloroform can be detected, and the nitro- glycerine weighed. It will, however, always be much too low. An actual analysis of a sample of gelatine dynamite gave the following result:—

Nitrocellulose (collodion) 3.819 per cent.
Nitro-glycerine 66.691 "
Wood-pulp 16.290 "
KNO_3 12.890 "

Na_2CO_3 *Nil.*
Water 0.340 "

This sample was probably intended to contain 30 per cent. of absorbing material to 70 per cent. of explosive substances. Many dynamites contain other substances than the above, such as paraffin, resin, sulphur, wood, coal-dust, charcoal, also mineral salts, such as carbonate of magnesia, chlorate of potash, &c. In these cases the above-described methods must of course be considerably modified. Paraffin, resin, and most of the sulphur will be found in the ether solution if present. The solution should be evaporated (and in this case the explosive should in the first case be treated with ether only, and not ether-alcohol), and the residue weighed, and then treated on the water bath with a solution of caustic soda. The resin goes into solution, and is separated by decantation from the residue, and precipitated by hydrochloric acid, and collected on a tarred filter (dried at 100° C.), and dried at 100° C. and weighed. The nitro-glycerine residue is treated with strong alcohol, decanted, and the residue of paraffin and sulphur washed with alcohol, dried, and weighed.

To separate the paraffin from the sulphur the residue is heated with a solution of ammonium sulphide. After cooling the paraffin collects as a crust upon the surface of the liquid, and by pricking a small hole through it with a glass rod the liquid underneath can be poured off, and the paraffin then washed with water, dried, and weighed. Sulphur is found by difference. Mr F.W. Smith (*Jour. Amer. Chem. Soc.*, 1901, 23 [8], 585-589) determines the sulphur in dynamite gelatine as follows: — About 2 grms. are warmed in a 100 c.c. silver crucible on the water bath with an alcoholic solution of sodium hydroxide, and where the nitro-glycerine is decomposed, the liquid is evaporated to dryness. The residue is fused with 40 grms. of KOH and 5 grms. of potassium nitrate, the mass dissolved in dilute acetic acid and filtered, and the sulphates precipitated in the usual way. If camphor is present, it can be extracted with bisulphide of carbon after the material has been treated with ether-alcohol. In that case the sulphur, paraffin, and resin will also be dissolved. The camphor being easily volatile, can be separated by evaporation. Let the weight of the extract, freed from ether-alcohol before treatment with bisulphide of carbon, equal A, and the weight of extract after treatment with CS_2 and evaporation of the same equal B; and weight of the residue which is left after evaporation of the CS_2 and the camphor in solution equal C, the percentage of camphor will be A - B - C. The residue C may contain traces of nitro-glycerine, resin, or sulphur.

Camphor may be separated from nitro-glycerine by means of CS_2. If the solution of camphor in nitro-glycerine be shaken with CS_2, the camphor and a little of the nitro-glycerine will dissolve. The bisulphide solution is decanted, or poured into a separating funnel and separated from the nitro-glycerine. The two solutions are then heated on the water bath to 20° C. and then to 60° C., and afterwards in a vacuum over $CaCl_2$ until the CS_2 has evaporated from them. The camphor evaporates, and leaves the small quantity of nitro-glycerine which had been dissolved with it. The other portion is the nitro-glycerine, now free from CS_2. The two are weighed and their weights added together, and equals the nitro-glycerine present. There is a loss of nitro-glycerine, it being partly evaporated along with the CS_2. Captain Hess has shown that it is equal to about 1.25 per cent. This quantity should therefore be added to that found by analysis. Morton Liebschutz, in a paper in the *Moniteur Scientifique* for January 1893, very rightly observes that the variety of dynamites manufactured is very great, all of them having a special composition which, good or bad, is sometimes of so complicated a nature that the determination of their elements is difficult.

The determination of nitro-glycerine in simple dynamite No. 1 is easy; but not so when the dynamite contains substances soluble in ether, such as sulphur, resin, paraffin, and naphthalene. After detailing at length the methods he employs, he concludes with the observation that the knowledge of the use of acetic acid—in which nitro-glycerine dissolves—for the determination of nitro-glycerine may be serviceable. Mr F.W. Smith[A] gives the following indirect method of determining nitro-glycerine in gelatine dynamite, &c. About 15 grms. of the sample are extracted with chloroform in a Soxhlet apparatus, and the loss in weight determined. In a second portion the moisture is determined. A third portion of about 2 grms. is macerated with ether in a small beaker, the ethereal extract filtered, and the process of extraction repeated three or four times.

The united filtrates are allowed to evaporate spontaneously, and the residue warmed gently on the water bath with 5 c.c. of ammonium sulphide solution, and 10 c.c. of alcohol until the nitro-glycerine is decomposed, after which about 250 c.c. of water and sufficient hydrochloric acid to render the liquid strongly acid, are added, and the liquid filtered. The precipitate is washed free from acid, and then washed through the filter with strong alcohol and chloroform into a weighed platinum dish, which is dried to constant weight at 50° C. The contents of the dish are now transferred to a silver crucible, and the sulphur determined. This amount of sulphur, deducted from the weight of the contents of the platinum dish, gives the quantity of substances soluble in chloroform with the exception of the nitro-glycerine, moisture, and sulphur. The amount of the former substances *plus* the moisture and sulphur, deducted from the total loss on extraction with chloroform, gives the quantity of nitroglycerine. Nitro-benzene may be detected, according to J. Marpurgo, in the following manner:—In a porcelain basin are placed two drops of liquid phenol, three drops of water, and a fragment of potash as large as a pea. The mixture is boiled, and the aqueous solution to be tested then added. On prolonged boiling nitrobenzene produces at the edge of the liquid a crimson ring, which on the addition of a solution of bleaching powder turns emerald-green. And nitro-glycerine in ether solution, by placing a few drops of the suspected solution, together with a drop or two of aniline, upon a watch-glass, evaporating off the ether, and then adding a drop of concentrated sulphuric acid to the residue, when, if nitroglycerine is present, the H_2SO_4 will strike a crimson colour, due to the action of the aniline sulphate upon the nitric acid liberated from the nitro- glycerine.

[Footnote A: "Notes on the Analysis of Explosives," *Jour. Amer. Chem. Soc.*, 1901, 23 [8], 585-589.]

~Tonite.~—The analysis of this explosive is a comparatively easy matter, and can be performed as follows:—Weigh out 10 grms., or a smaller quantity, and boil with water in a beaker, decanting the liquid four or five times, and filter. The aqueous solution will contain the nitrate of barium. Then put the residue on the filter, and wash two or three times with boiling water. Evaporate the filtrate to dryness in a platinum dish. Dry and weigh. This equals the $Ba(NO_3)_2$. If the sample is tonite No. 3, and contains di-nitro-benzol, treat first with ether to dissolve out this substance. Filter into a dish, and evaporate off the ether, and weigh the di-nitro-benzol, and afterwards treat residue with water as before. The residue is dried and weighed, and equals the gun-cotton present. It should then be treated with a solution of ether-alcohol in a conical flask, allowed to stand some three hours, then filtered through a weighed filter paper, dried at 40° C., and weighed. This will give the gun-cotton, and the difference between this last weight and the previous one will give the collodion-cotton. A portion of the residue containing both the gun-cotton and the soluble cotton can be tested in the nitrometer, and the nitrogen determined.

~Cordite.~ — This explosive consists of gun-cotton (with a little collodion-cotton in it as impurity), nitro-glycerine, and vaseline — the proportions being given as 30 per cent. nitro-glycerine, 65 per cent. gun-cotton, and 5 per cent. vaseline. Its analysis is performed by a modification of the method given for gelatines. Five grms. may be dissolved in ether-alcohol in a conical flask, allowed to stand all night, and then filtered through a linen filter. The residue is washed with a little ether, pressed, and dried at 40° C., and weighed. It equals the gun-cotton. The solution contains the nitro-glycerine, soluble cotton, and vaseline. The cotton is precipitated with chloroform, filtered off, dried, and weighed. The two ether-alcohol solutions are mixed, and carefully evaporated down in a platinum dish upon the water bath at a low temperature. The residue is afterwards treated with strong 80 per cent. acetic acid, which dissolves out any nitro-glycerine left in it. The nitro-glycerine is then obtained by difference, or the method suggested to me privately by Mr W.J. Williams may be used. The residue obtained by evaporation of the ether-alcohol solution, after weighing, is treated with alcoholic potash to decompose the nitro-glycerine, water is added and the alcohol evaporated off. Some ether is then added, and the mixture shaken, and the ether separated and evaporated, and the residue weighed as vaseline.

The moisture should, however, be determined by the method devised by Mr Arthur Marshall, F.I.C., of the Royal Gunpowder

Works, Waltham Abbey, which is carried out as follows:—The cordite or other explosive is prepared in the manner laid down for the Abel heat test, that is t say, it is ground in a small mill, and that portion is selected which passes through a sieve having holes of the size of No. 8 wire gauge, but not through one with holes No. 14 wire gauge.

[Illustration: FIG. 40.—MARSHALL'S APPARATUS FOR MOISTURE IN CORDITE.]

The form of apparatus used is shown in Fig. 40. It consists of an aluminium dish A, having the dimensions shown, and the glass cone B weighing not more than 30 grms. Five grms. of the cordite are weighed into the aluminium dish A. This is covered with the cone B, and the whole is accurately weighed, and is then placed upon a metal plate heated by steam from a water bath. It is left upon the bath until all the moisture has been driven off, then it is allowed to cool for about half-an-hour in a desiccator and is weighed. The loss in weight gives accurately the moisture of the sample. For cordite of the original composition, one hour's heating is sufficient to entirely drive off the moisture; for modified cordite containing 65 per cent. of gun-cotton, two hours is enough, provided that there be not more than 1.3 per cent. of moisture present.

If the proportion of nitro-glycerine be higher, a longer heating is necessary. The aluminium dish must not be shallower than shown in the figure, for if the distance between the substance and the edge of the glass cone be less than half an inch, some nitro-glycerine will be lost. Again, the sample must not be ground finer than stated, else some of the moisture will be lost in the grinding and sieving operations, and the result will be too low. In order to be able to drive off all the moisture in the times mentioned, it is essential that the glass cone shall not fit too closely on the aluminium dish, consequently the horizontal ledge round the top of the dish should be bent, so as to render it slightly untrue, and leave a clearance of about 0.02 inch in some places. If these few simple precautions be taken, the method will be found to be very accurate. Duplicate determinations do not differ more than 0.01 per cent.[A]

[Footnote A: "Determination of Moisture in Nitro-glycerine Explosives," by

A. Marshall, *Jour. Soc. Chem. Ind.*, Feb. 29, 1904, p. 154.]

~The Vaseline~ ($C_{16}H_{34}$), or petroleum jelly, used has a flash-point of 400° F. It must not contain more than 0.2 per cent. volatile matter when heated for 12 hours on the water bath, and should have a specific gravity of 0.87 at 100° F., and a melting point of 86° F. It is obtained during the distillation of petroleum, and consists mainly of the portions distilling above 200° C. It boils at about 278° C.

~Acetone~ ($CH_3CO.CH_3$), or dimethyl ketone, is formed when iso-propyl alcohol is oxidised with potassium bichromate and sulphuric. It is also produced in considerable quantities during the dry distillation of wood, and many other organic compounds. Crude wood spirit, which has been freed from acetic acid, consists in the main of a mixture of acetone and methyl- alcohol. The two substances may be roughly separated by the addition of calcium chloride, which combines with the methyl-alcohol. On subsequent distillation crude acetone passes over, and may be purified by conversion into the bisulphite compound.

Acetone is usually prepared, however, by the dry distillation of crude calcium or barium acetate.

$(CH_3.COO)_2Ca = CH_3.CO.CH_3 + CaCO_3.$

The distillate is fractionated, and the portion, boiling between 50° and 60° C., mixed with strong solution of sodium bisulphite. The crystalline cake of acetone sodium bisulphite, which separates on standing, is well pressed, to free it from impurities, decomposed by distillation with dilute sodium carbonate, and the aqueous distillate of pure acetone dehydrated over calcium chloride. Acetone is a colourless, mobile liquid of sp. gr. .792 at 20° C., it boils at 56.5° C., has a peculiar, pleasant, ethereal odour, and is mixible with water, alcohol, and ether in all proportions.

The acetone used in the manufacture of cordite should conform to the following specification:—

SPECIFICATION FOR ACETONE.

1. The acetone to be not more than 0.802 specific gravity at 60° F. When mixed with distilled water it must show no turbidity, and must leave no residue on evaporation at 212° F. On distillation, four-fifths by volume of the quantity taken must distil over at a temperature not exceeding 138° F. The residual matter left after this distillation must not contain, besides acetone, any ingredient that is not a bye-product incidental to the manufacture of acetone.

2. One c.c. of 0.10 per cent. solution in distilled water of pure permanganate of potash, added to 100 c.c. of the acetone, must retain its distinctive colour for not less than 30 minutes. This test should be made at a temperature of 60° F.

3. The acetone tested by the following method must not show more than 0.005 per cent. of acid, calculated to acetic acid:—

To 50 c.c. of the sample diluted with 50 c.c. of distilled water, with 2 c.c. of phenol-phthalein solution (1 gramme to 1,000 c.c. of 50 per cent. alcohol) added as an indicator, add from a burette N/100 sodium hydrate solution (1 c.c. 0.0006 gramme acetic acid), and calculate to acetic acid in the usual manner.

The water used for the dilution of the acetone must be carefully tested for acidity, and the pipettes used for measuring should not be blown out, as it would be possible thus to neutralise nearly 2 c.c. of the soda solution.

The presence of water in a sample of acetone may be detected by Schweitzer and Lungwitz's method (*Chem. Zeit.*, 1895, xix., p. 1384), which consists in shaking together equal volumes of acetone and petroleum ether (boiling point, 40° to 60° C.), when if present a separation of the liquid in layers will take place.

~Estimation of Acetone.~—Kebler (*Jour. Amer. Chem. Soc.*, 1897, 19, 316- 320) has improved Squibb's modification of Robineau and Rollins' method. The following solutions are required:—

(1.) A 6 per cent. solution of hydrochloric acid.

(2.) A decinormal solution of sodium thiosulphate.

(3.) Alkaline potassium iodide solution prepared by dissolving 250 grms. of potassium iodide in water, made up to a litre; dissolving 257 grms. of sodium hydroxide (by alcohol) in water, likewise made up to a litre. After allowing the latter to stand, 800 c.c. of the clear solution are added to the litre of KI.

(4.) Sodium hypochlorite solution: 100 grms. of bleaching powder (35 per cent.) are mixed with 400 c.c. of water: to this is added a hot solution of 120 grms. of crystallised sodium carbonate in 400 c.c. of water. After cooling, the clear liquid is decanted, the remainder filtered, and the filtrate made up to a litre; to each litre is added 25 c.c. of sodium hydroxide solution (sp. gr. 1.29).

(5.) An aqueous solution of the acetone, containing 1 or 2 per cent. of acetone.

(6.) Bicarbonated starch solution prepared by treating 0.125 grm. of starch with 5 c.c. of cold water, then adding 20 c.c. of boiling water, boiling a few minutes, cooling, and adding 2 grms. of sodium bicarbonate.

To 20 c.c. of the potassium iodide solution are added 10 c.c. of the diluted aqueous acetone, an excess of the sodium hypochlorite solution is then run in from a burette and well shaken for a minute. The mixture is then acidified with the hydrochloric acid solution, and while agitated, an excess of sodium thiosulphate solution is added, the mixture being afterwards allowed to stand a few minutes. The starch indicator is then added, and the excess of thiosulphate re-titrated. The relation of the sodium hypochlorite solution to the sodium thiosulphate being known, the percentage of acetone can be readily calculated.[A]

[Footnote A: See "The Testing of Acetone," Conroy, *Jour. Soc. Chem. Ind.*, 31st March 1900, vol. xix.]

Dr S.J.M. Auld has recently (*Jour. Chem. Soc.*, Feb. 15, 1906, vol. xxv.) worked out a volumetric method for the estimation of acetone, depending on the formation of bromoform, and its subsequent hydrolysis with alcoholic potash. The hydrolysis is probably expressed thus—

$$3CHBr_3 + 9KOH + C_2H_5OH = 3CO + C_2H_4 + 9KBr + 7H_2O$$

as it has been shown by Hermann and Long that exactly 3 volumes of carbon monoxide to 1 of ethylene are evolved. The residual potassium bromide is estimated by means of standard silver nitrate solution. Bromoform is specially suitable for this purpose for several reasons. It is very readily formed by the action of bromine and potash on acetone, and although very volatile in steam, it is not liable to loss due to its own evaporation. Further, its high molecular weight and large percentage of bromine conduce to accurate results, 58 grms. of acetone being responsible for the formation of 357 grms. of KBr. The method of carrying out the analysis is as follows:—

A known quantity of the solution to be tested, containing acetone to the extent of 0.1 to 0.2 grm., is pipetted into a 500 c.c. round-bottom flask, diluted with a little water, and mixed with 20 to 30 c.c. of a 10 per cent. solution of caustic potash. The flask is connected with a long reflex condenser, and is also fitted with a dropping funnel containing a solution of bromine in potassium bromide (200 grms. of Br and 250 grms. of KBr to 1 litre of water). The bromine solution is allowed to flow into the mixture until it has acquired a faint yellow tinge, the flask and its contents being then heated on the water bath at about 70° C. for half-an-hour. Bromine solution is added drop by drop until the slight coloration is permanent, excess of bromine being got rid of by boiling for a minute or two with a little more caustic potash. The mixture is then distilled until the distillate is free from bromoform, halogen being tested for in the usual manner. Water is added to the contents of the flask if necessary. It may be here observed that no acetone can be detected in the distillate by means of the mercuric oxide test, and free bromine is also absent. The condenser having been washed out with a little alcohol, in order to remove any traces of bromoform which may have collected, the distillate and washings are mixed with 50 c.c. of alcohol and sufficient solid caustic potash to make an approximately 10 per cent. solution. The mixture is then heated on the water bath under a reflux condenser until the bromoform is completely decomposed. This generally occupies about three-quarters of an hour. The liquid is allowed to cool, evaporated to smaller bulk if necessary, and exactly neutralised with dilute nitric acid. It is then diluted with water to 500 c.c., and an aliquot part titrated with N/10 silver nitrate solution, using potassium chromate as indicator; 240

parts of bromine correspond to 58 parts of acetone. The complete analysis can be performed in one and a half to two hours. It is imperative that the bromine used should be pure, as crude bromine frequently contains bromoform. The method is suitable for the estimation of acetone in wood-spirit, the spirit being diluted to 10 times its volume, and 5 c.c. of this solution employed for the determination. For example—

(1.) Three c.c. of a solution containing 9.61 per cent. acetone gave 1.7850 grm. KBr. Acetone found = 9.66 per cent.

(2.) Ten c.c. of a solution containing 0.96 per cent. acetone gave 0.5847 grm. KBr. Acetone found = 0.95 per cent.

~Nitro-Cotton.~—The first thing upon opening a case of wet cotton, or in receiving a sample from the "poacher," that requires to be determined is the percentage of water that it contains. It is best done by weighing out about 1,000 grms. upon a paper tray, which has been previously dried in the oven at 100° C. for some time, and become constant in weight. The trayful of cotton is then placed in a water oven, kept at 100° C., and dried as long as it loses water. The loss gives the percentage of water. It varies from 20 to 30 per cent. as a rule in "wet" cotton.

OUTLINE SCHEME FOR THE ANALYSIS OF NITRO-EXPLOSIVES

_____ | | | Exhaust dried substance with Anhydrous Ether in   Soxhlet's   Fat | | Extraction   Apparatus. |
|_____
_____ | | | *Solution*—Divide into two parts ~A.~ and ~B.~
|
|_____
_____ | | | ~A.~ | | | | Allow ether to evaporate spontaneously. Dry residue in vacuo over | | H_2SO_4 and weigh. Equals nitro-glycerine, resin, camphor, and | | paraffin. | | | | The nitro-glycerine in this residue may be decomposed by heating | | with a solution of alcoholic potash. Water may then be added, and the | | alcohol evaporated off on the water bath. From this solution the | | resin may be precipitated by HCl, filtered off, dried, and weighed. | | Solution containing the paraffin is treated with AmS

solution and | | heated. On cooling the paraffin separates, and may be separated. | | Residue may be shaken with CS_2 to remove camphor. |
|_____
_____| | | | ~B.~ | | | | Add phenol-phthalein and titrate with alcoholic potash, 1 c.c. normal | | KHO = .330 grm. *resin*, and add considerably more KHO. Evaporate, | | dissolve residue in water, shake with ether, and separate. |
|_____
_____| | | | *Ethereal Solution* evaporated leaves paraffin. |
|_____
_____| | | | *Aqueous Solution—* | | Add bromide, acidify with HCl, separate any resin and precipitate, | | filtrate with $BaCl_2$ $BaSO_4$ x   .1373   =   Sulphur. |
|_____
_____| | | | *Residue—* | | Dry, weigh, and exhaust with water preferably in Soxhlet. |
|_____
_____| | | | | *Solution—* | *Residue—* | | Contains metallic | Dry, weigh, and agitate an aliquot part with | | nitrates, chlorates, | with H_2SO_4 and Hg in nitrometer. If | | soluble carbonates, | nitro-cellulose is present, treat remainder of | | the sum of which | residue with ether-alcohol. | | (except $AmCO_3$)
|_____| | can be determined by | | | evaporating down at | *Solution—* | | 100° C. to dryness | Evaporate and weigh. Residue consists of | | and weighing. | soluble nitro-cellulose. | | Nitrates can be
_____		determined by				*Residue—*			Dry and weigh and determine hexa-nitro-			cellulose in nitrometer, if present. Exhaust			remainder with acetic ether.	
_____					
Solution—	*Residue—*			Hexa-nitro-cellulose	Dry and weigh, ignite			(Gun cotton).	and reweigh. Loss =				*Cellulose.*			|_____|						Residue consists of				sawdust, charcoal,				coal, chalk, guhr,				or mineral matter, &c.
_____	_____	_____
_____|

NOTE.—Camphor is found by difference. Sulphur is only partially soluble in ether. It is better, therefore, to extract some of the original substance with water, and treat residue with alcoholic KHO. Add bromide, acidify, and precipitate as BaSO.

~The Solubility Test.~—The object of this test is to ascertain, in the case of gun-cotton, the percentage of soluble (penta and lower nitrates) cotton that it contains, or in the case of soluble cotton, the quantity of gun-cotton. The method of procedure is as follows:— Five grms. of the sample which has been previously dried at 100° C., and afterwards exposed to the air for two hours, is transferred to a conical flask, and 250 c.c. ether-alcohol added (2 ether to 1 alcohol). The flask is then corked and allowed to digest, with repeated shaking, for two or three hours. The whole is then transferred to a linen filter, and when the solution has passed through the filter, is washed with a little ether, and pressed in a hand-screw press between folds of filter paper. The sample is then returned to the flask, and the previous treatment repeated, but it will be sufficient for it to digest for one hour the second time. The filter is then again pressed first gently by hand, then in the press, and afterwards opened up and the ether allowed to evaporate. The gun-cotton is then removed from the filter and transferred to a watch-glass, and dried in the water oven at 100° C. When dry it is exposed to the air for two hours and weighed. It equals the amount of gun-cotton and unconverted cotton in the 5 grms. The unconverted cotton must be determined in a separate 5 grms. and deducted.

The method of determining the soluble cotton now used in the Government laboratories is as follows:—Fifty grains of the nitro-cotton are dissolved in 150 c.c. of ether-alcohol, and allowed to stand, with frequent shakings, in a 200 c.c. stoppered measure for six hours; 75 c.c. of the clear solution are then drawn off by the aid of a pipette and evaporated in a dish on the water bath, and finally in the water oven at 120° F. (49° C.), until constant in weight. The weight found equals the quantity of soluble cotton in the 75 c.c., which, multiplied by 4, equals the percentage, thus: Suppose that 2.30 grains was the weight found, then

$(2.3 \times 150)/75 = 4.6$ in $50 = 9.20$ per cent.

A method for the determination of soluble nitro-cellulose in gun-cotton and smokeless powder has been published by K.B. Quinan (*Jour. Amer. Chem. Soc.*, 23 [4], 258). In this method about 1 grm. of the finely divided dry sample to be analysed is placed in an aluminium cup 1.9 inch in diameter and 4-1/8 inch deep. It is then covered and well stirred with 50 c.c. of alcohol, 100 c.c. of ether are then added, and the mixture is stirred for several minutes. After removing the stirrer, the cup is lightly covered with an aluminium lid, and is then placed in the steel cup of a centrifugal machine, which is gradually got up to a speed of 2,000 revolutions per minute, the total centrifugal force at the position occupied by the cups (which become horizontal when in rapid rotation) is about 450 lbs. They are rotated at the full speed for ten to twelve minutes, and the machine is then gradually stopped. By this time the whole of the insoluble matter will be at the bottom of the cup, and the supernatant solution will be clear. It is drawn off to within a quarter of an inch of the bottom (without disturbing the sediment), with the aid of a pipette.

Care must be taken that the solution thus withdrawn is perfectly clear. About 10 to 15 c.c. of colloid solution and a film of insoluble matter remain at the bottom of the cup; these are stirred up well, the stirrer is rinsed with ether-alcohol, about 50 c.c. of fresh ether-alcohol are added; the mixture is again treated in the centrifugal apparatus for about eight minutes; the whole washing process is then repeated until all soluble matter has been removed. This may require about seven or eight (or for samples with much insoluble matter ten or twelve or more) washings, but as the extraction proceeds, the period of rotation may be somewhat reduced. After extraction is completed, the insoluble matter is transferred to a Gooch crucible with the usual asbestos pad, dried at 100° C., and weighed. The residue may, if wished, be dried and weighed in the aluminium cup, but then it cannot be ignited. The whole time for an analysis exclusive of that required for drying, is from one to two hours — average time, 1-1/4 hour. The results are satisfactory both as to accuracy and rapidity. Acetone-soluble nitro-cellulose may be determined by the same method.

~The Unconverted or Non-nitrated Cotton.~ — However well the cotton has been nitrated, it is almost certain to contain a small quantity of non- nitrated or unconverted cotton. This can be determined thus: — Five grms. of the sample are boiled with a saturated solution of sodium sulphide, and then allowed to stand for forty-eight hours, and afterwards filtered or decanted, and again boiled with fresh solutions of sulphide, and again filtered, washed first with dilute HCl and then with water, dried, and weighed. The residue is the cellulose that was not nitrated, plus ash, &c. It should be ignited, and the weight of the ash deducted from the previous weight.

Acetone, and acetic-ether (ethyl-acetate) may also be used as solvents for the nitro-cellulose. Another process is to boil the gun-cotton, &c., in a solution of sodium stannate made by adding caustic soda to a solution of stannous chloride, until the precipitate first formed is just re-dissolved. This solution dissolves the cellulose nitrates, but does not affect the cellulose. Dr Lungé found the following process more satisfactory in the case of the more highly nitrated products: — The reagent is an alcoholic solution of sodium-ethylate prepared by dissolving 2 to 3 grms. of sodium in 100 c.c. of 95 per cent. alcohol, and mixing the filtered solution with 100 c.c. of acetone. It has no effect upon cellulose, but decomposes nitro-cellulose with the formation of a reddish brown compound, which is soluble in water. In the determination, 5 grms. of gun-cotton are heated to 40° or 50° C. on the water bath with 150 c.c. of the reagent, the liquid being shaken at intervals for twenty to thirty minutes; or the mixture may be allowed to stand for a few hours at the ordinary temperature. The brown-red solution is decanted from the undissolved residue, and the latter washed with alcohol and with water, by decantation, and then on the filter with hot water, to which a little hydrochloric acid is added for the final washings. For ordinary work this cellulose is dried immediately and weighed, but in exact determinations it is washed with alcohol, again treated with 50 c.c. of the reagent, and separated and washed as before. The cellulose thus obtained, gives no trace of gas in the nitrometer, and duplicate determinations agree within 0.1 to 0.2 per cent. when the weight of unchanged cellulose amounts to about 0.2 grm. Gun-cotton, which is completely soluble in acetone, contains only traces of cellulose, and when as much as 0.85 per cent. is present it does not dissolve entirely. This method is not applicable to the determination of cellulose in lower nitrated products, and Dr Lungé attributes this to the fact that these being prepared with less concentrated acid invariably contain oxy-cellulose.

~Alkalinity.~ — Five grms. of the air-dried and very finely divided sample are taken from the centre of the slabs or discs, and digested with about 20 c.c. of N/2 hydrochloric acid, and diluted with water to about 250 c.c., and shaken for about fifteen minutes. The liquid is then decanted, and washed with water until the washings no longer give an acid reaction. The solution, together with the washings, are titrated with N/4 sodium carbonate, using litmus as indicator.

~Ash and Inorganic Matter.~ — This is best determined by mixing 2 or 3 grms. of the nitro-cotton in a platinum crucible with shavings of paraffin, heating sufficiently to melt the paraffin, and then allowing the contents of the crucible to catch fire and burn away quietly. The temperature is then raised, and the carbonaceous residue incinerated, cooled, weighed, &c., and the percentage of ash calculated. Schjerning proceeds in the following way: — He takes 5 grms. of the nitro-cotton in a large platinum crucible, he then moistens it with a mixture of alcohol and ether, in which paraffin has been dissolved to saturation, and filtered and mixed with one-fourth of its volume of water. Some fragments of solid paraffin are then added, and the ether set on fire. Whilst this is in progress the crucible is kept in an oblique position, and is rotated so that the gun-cotton may absorb the paraffin uniformly. The partially charred residue is now rubbed down with a rounded glass rod, and the crucible is covered and heated for from fifteen to twenty minutes over the blow-pipe, the lid being occasionally removed. The residue is soon converted into ash, which is weighed, and then washed out into a porcelain basin and treated with hydrochloric acid heated to 90° C. The oxide of iron, alumina, lime, and magnesia are thus dissolved, and the silica remains as insoluble residue. The rest of the analysis is conducted according to the well-known methods of separation. The percentage of ash as a whole is generally all that is required.

~Examination of Nitrated Celluloses with Polarised Light.~ — Dr G. Lungé (*Jour. Amer. Chem. Soc.*, 1901, 23 [8], 527) has formed the following conclusions: — The most highly nitrated products appear blue in polarised light, but those containing between 13.9 and 13.0 per cent. of nitrogen cannot be distinguished from each other by polarisation. As the percentage of nitrogen rises, the blue colour becomes less intense, and here and there grey fibres can be observed, though not in proportion to the increase in the nitrogen. Below 12.4 per cent. of nitrogen, the fibres show a grey lustre, which usually appears yellow when the top light is cut off. Below 10 per cent. of nitrogen, the structure is invariably partially destroyed and no certain observations possible. It is only possible to distinguish with certainty, firstly any unchanged cellulose by its flashing up in variegated (rainbow) colours; and secondly, highly nitrated products (from 12.75 per cent. N upwards), by their flashing up less strongly in blue colours. The purple transition stage in the fibres containing over 11.28 per cent. of N (Chardonnet) was not observed by Dr Lungé.

~Determination of Nitrogen by Lungé Nitrometer.~ — The determination of the percentage of nitrogen in a sample of gun-cotton or collodion is perhaps of more value, and affords a better idea of its purity and composition, than any of the foregoing methods of examination, and taken in conjunction with the solubility test, it will generally give the analyst a very fair idea of the composition of his sample. If we regard gun-cotton as the hexa-nitro-cellulose, the theoretical amount of nitrogen required for the formula is 14.14 per cent., and in the same way for collodion-cotton, which consists of the lower nitrates, chiefly, however, of the penta- nitrate, the theoretical nitrogen is 12.75 per cent., so that if in a sample of nitro-cotton the nitrogen falls much lower than 14 per cent., it probably contains considerable quantities of the lower nitrates, and perhaps some non-nitrated cellulose as well $(C_{6}H_{10}O_{5})_{x}$, which of course would also lower the percentage of nitrogen.

The most expeditious method of determining the nitrogen in these nitro bodies is by the use of Lungé's nitrometer (Fig. 41), and the best way of working the process is as follows: — Weigh out with the greatest care 0.6 grm. of the previously dried substance in a small weighing bottle of about 15 c.c. capacity, and carefully add 10 c.c. of concentrated sulphuric acid from a pipette, and allow to stand until all the cotton is dissolved. The nitrometer should be of a capacity 150 to 200 c.c., and should contain a bulb of 100 c.c. capacity at the top, and should be fitted with a Greiner and Friederich's three-way tap. When the nitro-cotton has entirely dissolved to a clear solution, raise the pressure tube of the nitrometer so as to bring the mercury in the measuring tube close up to the tap. Open the tap in order to allow of the escape of any air bubbles, and clean the surface of the mercury and the inside of the cup with a small piece of filter paper. Now close the tap, and pour the solution of the nitro-cotton into the cup. Rinse out the bottle with 15 c.c. of sulphuric acid, contained in a pipette, pouring a little of the acid over the stopper of the weighing bottle in case some of the solution may be on it. Now lower the pressure tube a little, just enough to cause the solution to flow into the bulb of the measuring tube, when the tap is slightly opened. When the solution has run in almost to the end, turn off the tap, wash down the sides of the bottle, and add to the cup of the nitrometer; allow it to flow in as before, and then wash down the sides of the cup with 10 c.c. of sulphuric acid, adding little by little, and allowing each portion added to flow into the bulb of the nitrometer before adding the next portion. Great care is necessary to prevent air bubbles obtaining admission, and if the pressure tube is lowered too far, the acid will run with a rush and carry air along with it.

[Illustration: FIG. 41.—ORDINARY FORM OF LUNGÉ NITROMETER.]

The solution being all in the measuring tube, the pressure tube is again slightly raised, and the tube containing the nitro-cotton solution shaken for ten minutes with considerable violence. It is then replaced in the clamp, and the pressure relieved by lowering the pressure tube, and the whole apparatus allowed to stand for twenty minutes, in order to allow the gas evolved to assume the temperature of the room. A thermometer should be hung up close to the bulb of the measuring tube. At the end of the twenty minutes, the levels of the mercury in the pressure and measuring tubes are equalised, and the final adjustment obtained by slightly opening the tap on the measuring tube (very slightly), after first adding a little sulphuric acid to the cup, and observing whether the acid runs in or moves up. This must be done with very great care. When accurately adjusted, it should move neither way. Now read off the volume of the NO gas in cubic centimetres from the measuring tube. Read also the thermometer suspended near the bulb, and take the height of the barometer in millimetres. The calculation is very simple.

EXAMPLE—COLLODION-COTTON.

0.6[A] grm. taken. Reading on measuring tube = 114.6 c.c. NO. Barometer— 758 mm. Temperature—15° C.

[Footnote A: 0.5 grm. is enough in the case of gun-cotton.]

Since 1 c.c. NO = 0.6272 milligramme N, and correcting for temperature and pressure by the formula

760 x (1 + d^{2}) (d = .003665), for temperature 15° = 801.78,[A]

then

(114.6 x 100 x 750 x .6272)/(801.7 x. 6) = 11.22 per cent. nitrogen.

[Footnote A: See Table, page 244.]

The nitrogen in nitro-glycerine may of course be determined by the nitrometer, but in this case it is better to take a much smaller quantity of the substance. From 0.1 to 0.2 grm. is quite sufficient. This will give from 30 to 60 c.c. of gas, and therefore a measuring tube without a 100 c.c. bulb must be used.

EXAMPLE.

0.1048 grm. nitroglycerine taken gave 32.5 c.c. NO. Barometer, 761 mm. Temperature, 15° C.

Therefore,

(3.25 x 100 x 761 x .6272)/(801.78 x.1048) = 18.46 per cent. N. Theory = 18.50 per cent.

Professor Lungé has devised another form of nitrometer (Fig. 42), very useful in the nitrogen determination in explosives. It consists of a measuring tube, which is widened out in the middle to a bulb, and is graduated above and below into 1/10 c.c. The capacity of the whole apparatus is 130 c.c.; that of each portion of the tube being 30 c.c., and of the bulb 70 c.c. The upper portion of the graduated tube serves to measure small volumes of gas, whilst larger volumes are read off on the lower part.

[Illustration: FIG. 42. FIG. 43. SOME NEW FORMS OF NITROMETER.]

F.M. Horn (*Zeitschrift für angewandte Chemie*, 1892, p. 358) has devised a form of nitrometer (Fig. 43) which he has found especially useful in the examination of smokeless powders. The tap H is provided with a wide bore through which a weighed quantity of the powder is dropped bodily into the bulb K. From 4 to 5 c.c. of sulphuric acid which has been heated to 30° C. are then added through the funnel T, the tap H being immediately closed. When the powder has dissolved—a process which may be hastened by warming the bulb very carefully—the thick solution is drawn into the nitrometer tube N, and the bulb rinsed several times with fresh acid, after which operation the analysis is proceeded with in the usual way.

Dr Lungé's method of using a separate nitrometer in which to measure the NO gas evolved to the one in which the reaction has taken place, the gas being transferred from the one to the other by joining them by means of indiarubber tubing, and then driving the gas over by raising the pressure tube of the one containing the gas, the taps being open, I have found to be a great improvement.

1 c.c. NO gas at 0° and 760 mm.
Equals 0.6272 milligrammes (N) nitrogen.
" 1.343 " nitric oxide.
" 2.820 " (HNO_3) nitric acid.
" 3.805 " ($NaNO_3$) sodium nitrate.
" 4.523 " (KNO_3) potassium nitrate.

~Champion and Pellet's Method.~—This method is now very little used. It is based upon the fact that when nitro-cellulose is boiled with ferrous chloride and hydrochloric acid, all the nitrogen is disengaged as nitric oxide (NO). It is performed as follows:—A vacuum is made in a flask, fitted with a funnel tube, with a glass stopper on the tube; a delivery tube that can also be closed, and which dips under a solution of caustic soda contained in a trough, and the end placed under a graduated tube, also full of caustic soda. From 0.12 to 0.16 grm. cotton dissolved in 5 to 6 c.c. of sulphuric acid is allowed to flow into the flask, which contains the ferrous chloride and hydrochloric acid, and in which a vacuum has been formed by boiling, and then closing the taps. The solution is then heated, the taps on the delivery tube opened, and the end placed under the collecting tube, and the NO evolved collected. The NO gas is not evolved until the solution has become somewhat concentrated. Eder substituted a solution of ferrous sulphate in HCl for ferrous chloride. Care must be taken that the flask used is strong enough to stand the pressure, or it will burst.

The same chemists (*Compt. Rendus*, lxxxiii. 707) also devised the following method for determining the NO_2 in nitro-glycerine:— A known quantity of a solution of ferrous sulphate of previously ascertained reducing power is placed in a flask, acidified with hydrochloric acid, and its surface covered with a layer of petroleum oil. About .5 grm. of the nitro-glycerine is then introduced, and the flask heated on the water bath. When the sample is completely decomposed, the liquid is heated to boiling to remove nitric oxide, and the excess of ferrous sulphate ascertained by titration with standard permanganate; 56 of iron (Fe) oxidised by the sample correspond to 23 of NO_2 in the sample of nitro-glycerine.

~The Schultze-Tieman Method~ for determining nitrogen in nitro-explosives, especially nitro-cellulose and nitro-glycerine.—The figure (No. 44) shows the general arrangement of the apparatus. I am indebted for the following description of the method of working it to my friend, Mr William Bate, of Hayle. To fill the apparatus with the soda solution, the gas burette is put on the indiarubber stopper of basin W, and firmly clamped down. Then the taps A and C are opened, and B closed. When the burette is filled with soda solution half-way up the funnel Y, A and C are closed, and B opened. The arrows show the inlet and outlet for the cooling water that is kept running through the water jacket round the nitrometer tube. To collect the gas, raise the nitrometer off the rubber stopper, and place the gas tube from the decomposition apparatus in the glass dish W and under the opening of the nitrometer.

[Illustration: Fig. 44. SCHULTZE-TIEMAN APPARATUS.]

For the estimation of nitrogen in nitro-cellulose take .5 to .65 grm., and place in the decomposition flask *f* (Fig. 45), washing in with about 25 c.c. of water by alternately opening clips D and E. The air in the flask is driven out by boiling, whilst the air is shut off by the tube *i* dipping into the basin W, which is filled with the soda lye, and tube K is placed in the test tube R, which contains a few c.c. of water. As soon as all the air is completely driven out, clips D and E are closed, and the gas jet is taken away. (This flask must be a strong one, or it will burst.) Into test tube R, 25 c.c. of concentrated solution of protochloride of iron and 10 to 15 c.c. concentrated hydrochloric acid are poured, which are sucked up into the developing flask *f* by opening clip E, air being carefully kept from entering. The clip E is now closed, and tube *i* is put underneath the burette, and the development of NO gas is commenced by heating the contents of the flask *f*. When the pressure of the gas in the flask has become greater than the pressure of the atmosphere, the connecting tube begins to swell at *i*, whereupon clip D is opened, and the boiling continued with frequent shaking of the bulb, until no more nitrous gas bubbles rise up into the soda lye, the distilling over of the HCl causes a crackling noise, the clip D is closed, and E opened. The burette is again put hermetically on the indiarubber stopper in basin W, and the apparatus is left to cool until the water discharged through P shows the same temperature as the water flowing through (into the cooling jacket) Z. If the level of the soda solution in the tube X is now put on exactly the same level as that in the burette by lowering or elevating the tube X as required, the volume of NO obtained in c.c. can be read off within 1/10 c.c., and the percentage of nitrogen calculated by the usual formula.

[Illustration: FIG. 45.—Decomposition Flask for Schultze-Tieman Method.]

The solution of protochloride of iron is obtained by dissolving iron nails, &c., in concentrated HCl, the iron being in excess. When the development of hydrogen ceases, it is necessary to filter warm through a paper filter, and acidify filtrate with a few drops of HCl. The soda solution used has a sp. gr. of 1.210 to 1.260; equals 25° to 30° B. The nitro-cellulose is dried in quantities of 2 grms. at 70° C. during eight to ten hours, and then three hours in an exiccator over H_2SO_4. The results obtained with this apparatus are very accurate. The reaction is founded upon that of MM. Champion and Pellet's method.

~The Kjeldahl Method of Determining Nitrogen.~—This method, which has been so largely used by analysts for the determination of nitrogen in organic bodies, more especially perhaps in manures, was proposed by J. Kjeldahl,[A] of the Carlsberg Laboratory of Copenhagen. It was afterwards modified by Jodlbauer, of Munich,[B] and applied to the analysis of nitro- explosives by M. Chenel, of the Laboratoire Centrale des Poudres, whose method of procedure is as follows:—0.5 grm. of the finely powdered substance is digested in the cold with a solution of 1.2 grm. of phenol and 0.4 grm. phosphoric anhydride in 30 c.c. of sulphuric acid. The mixture is kept well shaken until the solution is complete. From 3 to 4 grms. of zinc-dust is then cautiously and gradually added, the temperature of the mass being kept down until complete reduction has been effected. Finally, 0.7 grm. of mercury is added, and the process continued in the usual way, according to Kjeldahl; that is, the liquid is distilled until all the ammonia has passed over, and is absorbed in the standard acid. The distillate is then titrated with standard ammonia.

[Footnote A: J. Kjeldahl, *Zeitschrift Anal. Chem.*, 1883, xxii., p. 366.]

[Footnote B: Jodlbauer, *Chemisches Centralblatt*, 1886, pp. 434-484. See also *Arms and Explosives*, 1893, p. 87.]

The NO_2 group is at the moment of solution fixed upon the phenol with the production of mono-nitro-phenol, which is afterwards reduced by the action of the zinc-dust into the amido derivative. During the subsequent combustion, the nitrogen of the amido-phenol becomes fixed in the state of ammonia. M. Chenel is perfectly satisfied with the results obtained, but he points out that the success of the operation depends upon the complete conversion of the phenol into the mono-nitro derivatives. This takes place whenever the organic compound forms a *clear solution* in the cold sulphuric acid mixture. Substances like collodion or gun-cotton must be very finely divided for successful treatment. The following table shows some of the results obtained by M. Chenel:—

Substances Analysed.	Total Nitrogen.				
	Calculated.	Found.			
Saltpetre (KNO_3)	13.86	13.91	13.82	13.73	13.96
Ammonium nitrate	35.00	35.31	34.90	34.96	
Barium nitrate	10.72	10.67	10.62		
Nitro-glycerol	18.50	18.45			
Di-nitro-benzol[A]	16.67	16.78	16.57		
Para-nitro-phenol	10.07	10.03			
Picric

| acid[A] | 18.34 | 18.42 | | | | 18.43 | | | Ammonium picrate | 22.76 | 22.63 | | | | 22.67 | | Di-nitro-ortho-cresol | 14.14 | 14.10 | | | | 13.98 | | Tri-nitro-meta-cresol | 17.28 | 17.57 | | | | 17.27 |

[Footnote A: Dr. Bernard Dyer obtained 18.39 per cent. for picric acid and 16.54 per cent. for di-nitro-benzol. — *Jour. Chem. Soc.*, Aug. 1895.]

When Chenel endeavoured to apply Jodlbauer's modification of Kjeldahl's process to the examination of the tri- and tetra-nitrated naphthalenes, he found that good results were not obtainable, because these compounds do not dissolve completely in the cold sulphuric acid. It may, however, be used if they are previously converted into the naphthylamines, according to the plan proposed by D'Aguiar and Lautemann (*Bull. Soc. Chim.*, vol. iii., new series, p. 256). This is rapidly effected as follows: — Twelve grms. of iodine are gradually added to a solution of 2 grms. of phosphorus in about 15 or 20 c.c. of bisulphide of carbon, this solution being contained in a flask of 250 c.c. capacity. The flask and its contents are heated on the water bath at 100° C. with constant attention, until the last traces of the carbon bisulphide have distilled away. It is then cooled, and the iodide of phosphorus is detached from the sides of the flask by shaking, but not expelled. The next step is to add about 0.5 to 0.6 grm. of the substance that is to be analysed, after which 8 grms. of water are introduced, and the flask is agitated gently two or three times. As soon as the reaction becomes lively, the contents of the flask are well shaken. It is usually finished about one minute after the addition of the water. The flask is now cooled, and 25 c.c. of sulphuric acid, together with 0.7 grm. of mercury, are gradually added; hydriodic acid (HI) forms, and the temperature of the flask must be raised sufficiently to expel it. The remaining part of the operation is as in the ordinary Kjeldahl process.

M. Chenel has found this process the best for the analysis of the nitro- naphthalenes, and for impervious substances like collodion or gun-cotton. Personally, I have never been able to obtain satisfactory results with this process in the analysis of nitro-cellulose, and I am of opinion that the process does not possess any advantage over the nitrometer method, at any rate for the analysis of gun-cotton.

Table giving the Percentages of Nitrogen and Oxide of Nitrogen in Various
Substances used in or as Explosives:

Name FORMULÆ NITROGEN NO_2
                        per cent. per cent.

Nitroglycerine $C_3H_5(ONO_2)_3$ 18.50 = 60.70
Hexa-nitro-cellulose $C_{12}H_{14}O_4(ONO_2)_6$ 14.14 = 46.42
Penta-nitro-cellulose $C_6H_8O_5(ONO_2)_5$ 11.11 = 36.50
Nitro-benzene $C_6H_5NO_2$ 11.38 = 37.39
Di-nitro-benzene $C_6H_4(NO_2)_2$ 16.67 = 54.77
Tri-nitro-benzene $C_6H_3(NO_2)_3$ 19.24 = 63.22
Nitro-toluene $C_7H_7NO_2$ 10.21 = 33.49
Nitro-naphthalene $C_{10}H_7NO_2$ 8.09 = 26.53
Di-nitro-naphthalene $C_{10}H_6(NO_2)_2$ 12.84 = 42.12
Nitro-mannite $C_6H_7(NO_3)_6$ 23.59 = 77.37
Nitro-starch $C_6H_8O_4(HNO_3)$ 6.76 = 22.18
Picric acid
  (Tri-nitro-phenol) $C_6H_2OH(NO_2)_3$ 18.34 = 60.15
Chloro-nitro-benzene $C_6H_3Cl(NO_2)_2$ 13.82 = 45.43
Ammonium nitrate NH_4NO_3 35.00 =
Sodium nitrate $NaNO_3$ 16.47 =
Potassium nitrate KNO_3 13.86 =
Nitric acid HNO_3 22.22 =
Barium nitrate $Ba(NO_3)_2$ 10.72 =

~Analysis of Celluloid.~—The finely divided celluloid is well stirred, by means of a platinum wire, with concentrated sulphuric acid in the cup of a Lungé nitrometer, and when dissolved the nitrogen determined in the solution in the usual way. To prevent interference from camphor, the following treatment is suggested by H. Zaunschirm (*Chem. Zeit.*, xiv., 905). Dissolve a weighed quantity of the celluloid in a mixture of ether- alcohol, mixed with a weighed quantity of washed and ignited asbestos, or pumice-stone, dry, and disintegrate the mass, and afterwards extract the camphor with chloroform, dry, and weigh: then extract with absolute methyl-alcohol, evaporate, weigh, and examine the nitro-cellulose in the nitrometer.

~Picric Acid and Picrates.~ — Picric acid is soluble in hot water, and to the extent of 1 part in 100 in cold water, also in ether, chloroform, glycerine, 10 per cent. soda solution, alcohol, amylic alcohol, carbon bisulphide, benzene, and petroleum. If a solution of picric acid be boiled with a strong solution of potassium cyanide, a deep red liquid is produced, owing to the formation of potassium iso-purpurate, which crystallises in small reddish-brown plates with a beetle-green lustre. This, by reaction with ammonium chloride, gives ammonium iso-purpurate ($NH_4C_8H_4N_5O_6$), or artificial murexide, which dies silk and wool a beautiful red colour. On adding barium chloride to either of the above salts, a vermilion-red precipitate was formed, consisting of barium iso-purpurate. With ammonio-sulphate of copper, solutions of picric acid give a bright green precipitate. Mr A.H. Allen gives the following methods for the assay of commercial picric acid, in his "Commercial Organic Analysis": —

~Resinous and Tarry matters~ are not unfrequently present. They are left insoluble on dissolving the sample in boiling water. The separation is more perfect if the hot solution be exactly neutralised by caustic soda.

~Sulphuric Acid, Hydrochloric Acid, and Oxalic Acid~, and their salts are detected by adding to the filtered aqueous solution of the sample solutions of the picrates of barium, silver, and calcium. These salts are readily made by boiling picric acid with the carbonates of the respective metals and filtering: other soluble salts of these methods may be substituted for the picrates, but they are less satisfactory.

~Nitric Acid~ may be detected by the red fumes evolved on warming the sample with copper turnings.

~Inorganic Impurities and Picrates of Potash and Sodium~, &c., leave residues on cautious ignition.

~General Impurities and Adulterations~ may be detected and determined by shaking 1 grm. of the sample of acid in a graduated tube with 25 c.c. of ether, the pure acid dissolves, while any oxalic acid, nitrates, picrates, boric acid, alum, sugar, &c., will be left insoluble, and after removal of the ethereal liquid, may be readily identified and determined. For the detection and determination of water and of oxalic acid, 50 c.c. of warm benzene may be advantageously substituted for ether. Sugar may be separated from the other impurities by treating the residue insoluble in ether or benzene with rectified spirit, in which sugar and boric acid alone will dissolve. If boric acid be present, the alcoholic solution will burn with a green flame. Mono- and di-nitrophenic acids lower the melting point (122° C). Their calcium salts are less soluble than the picrate, and may be approximately separated from it by fractional crystallisation, or by precipitating the hot saturated solution of the sample with excess of lime water. Picric acid may be determined by extracting the acidulated aqueous solution by agitation with ether or benzene, and subsequently removing and evaporating off the solvent. It may also be precipitated as the potassium salt.

~Potassium Picrate~ [$KC_6H_2(NO_2)_3O$]. When a strong solution of picric acid is neutralised by carbonate of potash, this salt is thrown down in yellow crystalline needles, which require 260 parts of cold or 14 parts of hot water for their solution. In alcohol it is much less soluble.

~Ammonium Picrate~ is more soluble in water than the above, and sodium picrate is readily soluble in water, but nearly insoluble in solution of sodium carbonate.

~Picrates of the Alkaloids.~—Picric acid forms insoluble salts with many of the alkaloids, and picric acid may be determined in the following manner:—To the solution of picric acid, or a picrate, add a solution of sulphate of cinchonine acidulated with H_2SO_4. The precipitated picrate of cinchonine [$C_{20}H_{24}N_2O(C_6H_2N_3O_7)_2$] is washed with cold water, rinsed off the filter into a porcelain crucible or dish, the water evaporated on the water bath, and the residual salt weighed. Its weight, multiplied by .6123, gives the quantity of picric acid in the sample taken.

~Analysis of Glycerine.~[A] Glycerine that is to be used for the manufacture of nitro-glycerine should have a minimum specific gravity of 1.261 at 15° C. This can be determined, either by the aid of a Sartorius specific gravity balance, or by using an ordinary specific gravity bottle. One of 10 or 25 c.c. capacity is very convenient.

[Footnote A: See also Sulman and Berry, *Analyst*, xi., 12-34, and Allen's "Commercial Organic Analysis," vol. ii., part i.]

~Residue~[A] left upon evaporation should not be more than 0.25 per cent. To determine this, take 25 grms. of the glycerine, and evaporate it at a temperature of about 160° C. in a platinum basin, and finish in an air bath. Weigh until constant weight is obtained. Afterwards incinerate over a bunsen burner, and weigh the ash.

[Footnote A: Organic matter up to .6 per cent. is not always prejudicial to the nitrating quantities of a glycerine.]

~Silver Test.~ A portion of the sample of glycerine to be tested should be put in a small weighing bottle, and a quarter of its bulk of N/10 silver nitrate solution added to it, then shake it, and place in a dark cupboard for fifteen minutes. It must be pronounced bad if it becomes black or dark brown within that time (acrolein, formic, and butyric acids).

The German official test for glycerine for pharmaceutical purposes is much more stringent, 1 c.c. of glycerine heated to boiling with 1 c.c. of ammonia solution and three drops of silver nitrate solution must give neither colour or precipitate within five minutes.

~Nitration.~ Fifty grms. of the glycerine are poured from a beaker into a mixture of concentrated nitric acid (specific gravity 1.53) and sulphuric acid (1.84), mixed in the proportions of 3 HNO_3 to 5 H_2SO_4 (about 400 c.c. of mixed acids). The mixed acids should be put into a rather large beaker, and held in the right hand in a basin of water, and the glycerine slowly poured into them from a smaller one held in the left. A constant rotatory motion should be given to the beaker in which the nitration is performed. When all the glycerine has been added, and the mixture has been shaken for a few minutes longer, it is poured into a separator, and allowed to stand for some time. It should, if the glycerine is a good one, have separated from the mixed acids in ten minutes, and the line of demarcation between the nitro-glycerine and the acid should be clear and sharp, neither should there be any white flocculent matter suspended in the liquid. The excess of acids is now drawn off, and the nitro-glycerine shaken once or twice with a warm solution of carbonate of soda, and afterwards with water alone. The nitro-glycerine is then drawn off into a weighed beaker, the surface dried with a piece of filter paper, and weighed; 100 parts of a good glycerine should yield about 230 of nitro-glycerine. A quicker method is to take only 10 c.c. of the glycerine, of which the specific gravity is already known, nitrate as before, and pour into a burette, read off the volume of nitro-glycerine in c.c. and multiply them by 1.6 (the specific gravity of nitro-glycerine), thus: 10 grms. gave 14.5 c.c. nitro-glycerine, and 14.5 x 1.6 = 23.2 grms., therefore 100 would give 232 grms. nitro-glycerine. The points to be noted in the nitration of a sample of glycerine are: the separation should be sharp, and within half an hour or less, and there should be no white flocculent matter formed, especially when the carbonate of soda solution is added.

~Total Acid Equivalent.~ Mr G.E. Barton (*Jour. Amer. Chem. Soc.*, 1895) proposes to determine thus: 100 c.c. of glycerine are diluted to 300 c.c. in a beaker, a few drops of a 1 per cent. solution of phenolphthalein and 10 c.c. of normal caustic soda solution are added; after boiling, the liquid is titrated with normal hydrochloric acid (fatty acids are thus indicated and roughly determined).

~Neutrality.~ The same chemist determines the neutrality of glycerine thus: 50 c.c. of glycerine mixed with 100 c.c. of water and a few drops of alcoholic phenolphthalein[A] are titrated with hydrochloric acid or sodium hydroxide; not more than 0.3 c.c. normal hydrochloric acid or normal soda solution should be required to render the sample neutral; raw glycerines contain from .5 to 1.0 per cent. of sodium carbonate.

[Footnote A: Sulman and Berry prefer litmus as indicator.]

~Determination of Free Fatty Acids.~ A weighed quantity of the glycerine is shaken up with some neutral ether in a separating funnel, the glycerine allowed to settle, drawn off, and the ether washed with three separate lots of water. The water must have been recently boiled, and be quite free from CO_2. All the free fatty acid is now in the ether, and no other soluble acid. A drop of phenolphthalein is now added, a little water, and the acidity determined by titration with deci-normal baryta solution, and the baryta solution taken calculated as oleic acid.

~Combined Fatty Acid.~ About 30 grms. of the glycerine are placed in a flask, and to it is added about half a grm. of caustic soda in solution. The mixture is heated for ten minutes at 150° C. After cooling some pure ether is added to it, and enough dilute H_2SO_4 to render it distinctly acid. It is well shaken. All the fatty acids go into the ether. The aqueous solution is then removed, and the ether well washed to remove all H_2SO_4. After the addition of phenolphthalein the acid is titrated, and the amount used calculated into oleic acid. From this total amount of fatty acids the free fatty acid is deducted, and the quantity of combined fatty acids thus obtained.

~Impurities.~ The following impurities may be found in bad samples of glycerine:—Lead, arsenic, lime, chlorine, sulphuric acid, thio-sulphates, sulphides, cyanogen compounds, organic acids (especially oleic acid and fatty acids[A]), rosin products, and other organic bodies. It is also said to be adulterated with sugar and glucose dextrine. Traces of sulphuric acid and arsenic may be allowed, also very small traces indeed of lime and chlorine.

[Footnote A: These substances often cause trouble in nitrating, white flocculent matter being formed during the process of washing.]

The organic acids, formic and butyric acids may be detected by heating a sample of the glycerine in a test tube with alcohol and sulphuric acid, when, if present, compound ethers, such as ethylic formate and butyrate, the former smelling like peaches and the latter of pine-apple, will be formed.

~Oleic Acid~, if present in large quantity, will come down upon diluting the sample with water, but smaller quantities may be detected by passing a current of nitrogen peroxide, N_2O_4 (obtained by heating lead nitrate), through the diluted sample, when a white flocculent precipitate of elaidic acid, which is less soluble than oleic acid, will be thrown down. By agitating glycerol with chloroform, fatty acids, rosin oil, and some other impurities are dissolved, while certain others form a turbid layer between the chloroform and the supernatant liquid. On separating the chloroform and evaporating it to dryness, a residue is obtained which may be further examined.

~Sodium Chloride~ can be determined in 100 c.c. of the glycerine by adding a little water, neutralised with sodium carbonate, and then titrated with a deci-normal solution of silver nitrate, using potassium chromate as indicator.

~Organic Impurities~ of various kinds occur in crude glycerine, and are mostly objectionable. Their sum may be determined with fair accuracy by Sulman and Berry's method: 50 grms. of the sample are diluted with twice its measure of water, carefully neutralised with acetic acid, and warmed to expel carbonic acid; when cold, a solution of basic lead acetate is added in slight but distinct excess, and the mixture well agitated. The formation of an abundant precipitate, which rapidly subsides, is an indication of considerable impurity in the sample. To ascertain its amount, the precipitate is first washed by decantation, and then collected on a tared, or preferably a double counter-poised filter, where it is further washed, dried at 100° to 105° C., and weighed. The precipitate and filter paper are then ignited separately in porcelain, at a low red heat, the residues moistened with a few drops of nitric acid and reignited; the weight of the lead oxide deducted from that of the original precipitate gives the weight of the organic matter precipitated by the lead. Raw glycerines contain from 0.5 to 1.0 per cent.

~Albuminous Matters.~ An approximate determination of the albuminous matters may be made by precipitating with basic lead acetate as already described, and determining the nitrogen by the Kjeldahl method; the nitrogen multiplied by 6.25 gives the amount of albuminous matter in the precipitate.

~The Determination of Glycerine.~ The acetin method of Benedikt and Canton depends upon the conversion of glycerine into triacetin, and the saponification of the latter, and reduces the estimation of glycerine to an acidmetric method. About 1.5 grm. of crude glycerine is heated to boiling with 7 grms. of acetic anhydride, and 3 to 4 grms. of anhydrous sodium acetate, under an upright condenser for one and a half hours. After cooling, 50 c.c. of water are added, and the mixture heated until all the triacetin has dissolved. The liquid is then filtered into a large flask, the residue on the filter is well washed with water, the filtrate quite cooled, phenolphthalein is added and the fluid exactly neutralised with a dilute (2 to 3 per cent.) solution of alkali. Twenty-five c.c. of a 10 per cent. caustic soda solution, which must be accurately standardised upon normal acid, are then pipetted into the liquid, which is heated to boiling for ten minutes to saponify the triacetin, and the excess of alkali is then titrated back with normal acid. One c.c. of normal acid corresponds to .03067 grm. of glycerine.

~Precautions.~—The heating must be done with a reflux condenser, the triacetin being somewhat volatile. The sodium acetate used must be quite anhydrous, or the conversion of the glycerine to triacetyl is imperfect. Triacetin in contact with water gradually decomposes. After acetylation is complete, therefore, the operations must be conducted as rapidly as possible. It is necessary to neutralise the free acetic acid very cautiously, and with rapid agitation, so that the alkali may not be locally in excess.

~The Lead Oxide Method.~—Two grms. of sample are mixed with about 40 grms. of pure litharge, and heated in an air bath to 130° C. until the weight becomes constant, care being taken that the litharge is free from such lead compounds and other substances as might injuriously affect the results, and that the heating of the mixture takes place in an air bath free from carbonic acid. The increase in weight in the litharge, minus the weight of substance not volatilisable from 2 grms. of glycerine at 160° C., multiplied by the factor 1.243, is taken as the weight of glycerine in the 2 grms. of sample. The glycerine must be fairly pure, and free from resinous substances and SO_3, to give good results by this process.

~Analysis of the "Waste Acids" from the Manufacture of Nitro-Glycerine or Gun-Cotton.~ Determine the specific gravity by the specific gravity bottle or hydrometer, and the oxides of nitrogen by the permanganate method described under nitro-glycerine. Now determine the total acidity of the mixture by means of a tenth normal solution of sodium hydrate, and calculate it as nitric acid (HNO_3), then determine the nitric acid by means of Lungé nitrometer, and subtract percentage found from total acidity, and calculate the difference into sulphuric acid, thus:—

Total acidity equals 97.46 per cent.—11.07 per cent. HNO_3 = 86.39 per cent., then (86.39 x 49)/63 = 67.20 per cent. H_2SO_4.

Then analysis of sample will be:—

Sulphuric acid = 67.20 per cent. |
Nitric acid = 11.07 " |- Specific gravity = 1.7075.
Water = 12.73 " |

This method is accurate enough for general use in the nitric acid factory. The acid mixture may be taken by volume for determining nitric oxide in nitrometer. Two c.c. is a convenient quantity in the above case, then 2 x 1.7075 (specific gravity) = 3.414 grms. taken, gave 145 c.c. NO (barometer = 748 mm, and temperature = 15°C.) equals 134.9 c.c. (corr.) and as 1 c.c. NO = .0282 grm. HNO_3 135 x .0282 = .378 grm. = 11.07 per cent. nitric acid.

~Sodium Nitrate.~ Determine moisture and chlorine by the usual methods, and the total, $NaNO_3$, by means of nitrometer—0.45 grm. is a very convenient quantity to work on (gives about 123 c.c. gas); grind very fine, and dissolve in a very little hot water in the cup of the nitrometer; use about 15 c.c. concentrated H_2SO_4. One cubic cent. of NO equals .003805 grm. of $NaNO_3$. The insoluble matter, both organic and inorganic, should also be determined, also sulphate of soda and lime tested for.

~Analysis of Mercury Fulminate (Divers and Kawakita's Method).~—A weighed quantity of mercury fulminate is added to excess, but measured quantity of fuming hydrochloric acid contained in a retort connected with a receiver holding water. After heating for some time, the contents of the retort and receiver are mixed and diluted, and the mercury is precipitated by hydrogen sulphide. By warming and exposure to the air in open vessels the hydrogen sulphide is for the most part dissipated. The solution is then titrated with potassium hydroxide (KOH), as well as another quantity of hydrochloric acid, equal to that used with the fulminate. As the mercury chloride is reconverted into hydrochloric acid by the hydrogen sulphide, and as the hydroxylamine does not neutralise to litmus the hydrochloric acid combined with it, there is an equal amount of hydrochloric acid free or available in the two solutions.

Any excess of acid in the one which has received the fulminate will therefore be due to the formic acid generated from the fulminate. Dr. Divers and M. Kawakita, working by this method, have obtained 31.31 per cent. formic acid, instead of 32.40 required by theory. (*Jour. Chem. Soc.*, p. 17, 1884.)

Divers and Kawakita proceed thus: 2.351 grms. dissolved, as already described, in HCl, and afterwards diluted, gave mercury sulphide equal to 70.40 per cent. mercury. The same solution, after removal of mercury, titrated by iodine for hydroxylamine, gave nitrogen equal to 9.85 per cent., and when evaporated with hydroxyl ammonium chloride equal to 9.55 per cent. A solution of 2.6665 grms. fulminate in HCl of known amount, after removal of mercury by hydrogen sulphide, gave by titration with potassium hydrate, formic acid equal to 8.17 per cent. of carbon. Collecting and comparing with calculation from formula we get—

Calc. I. II. III.

Mercury 70.42 70.40 ... ...
Nitrogen 9.86 9.85 9.55 ...
Carbon 8.45 ... ... 8.17
Oxygen 11.27 ... ...

100.00

~The Analysis of Cap Composition.~—Messrs F.W. Jones and F.A. Willcox (*Chem. News*, Dec. 11, 1896) have proposed the following process for the analysis of this substance:—Cap composition usually consists of the ingredients—potassium chlorate, antimony sulphide, and mercury fulminate, and to estimate these substances in the presence of each other by ordinary analytical methods is a difficult process. Since the separation of antimony sulphide and mercury fulminate in the presence of potassium chlorate necessitates the treatment of the mixture with hydrochloric acid, and this produces an evolution of hydrogen sulphide from the sulphide, and a consequent precipitation of sulphur; and potassium chlorate cannot be separated from the other ingredients by treatment with water, owing to the appreciable solubility of mercury fulminate in cold water.

In the course of some experiments on the solubility of mercury fulminate Messrs Jones and Willcox observed that this body was readily soluble in acetone and other ethereal solvents when they were saturated with ammonia gas, and that chlorate of potash and sulphide of antimony were insoluble in pure acetone saturated with ammonia; these observations at once afforded a simple method of separating the three ingredients of cap composition. By employing this solution of acetone and ammonia an analysis can be made in a comparatively short time, and yields results of sufficient accuracy for all technical purposes. The following are the details of the process:—

A tared filter paper is placed in a funnel to the neck of which has been fitted a piece of rubber tubing provided with a clip. The paper is moistened with a solution of acetone and ammonia, the cap composition is weighed off directly on to the filter paper and is then covered with the solution of acetone and ammonia and allowed to stand thirty-four hours. It is then washed repeatedly with the same solution until the washings give no coloration with ammonium sulphide, and afterwards washed with acetone until washings give no residue on evaporation dried and weighed. The paper is again put in the funnel and washed with water until free from potassium chlorate, dried and weighed.

If c = weight of composition taken,
d = " " filter paper,
a = " after first extraction,
b = " " second extraction,
  then $c+d-a$ = weight of fulminate,
$c+d-a-b$ = " " $KClO_3$,
$b-d$ = " " sulphide of antimony.

The composition should be finely ground in an agate mortar.

The results of the analysis by this method of two mixtures of known composition are given below—

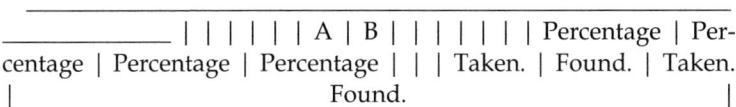

Antimony Sulphide	36.47	36.25	37.34	37.22
Potassium Chlorate	33.25	33.71	46.03	46.43
Mercury Fulminate	30.27	30.02	16.61	16.34

Dr. H.W. Brownsdon's (*Jour. Soc. Chem. Ind.*, xxiv., April 1905) process is as follows:—The cap composition is removed by squeezing the cap with pliers, while held over a porcelain basin of about 200 c.c. capacity, and removing the loosened foil and broken composition by means of a pointed wooden chip. Composition adhering to the shell or foil is loosened by alcohol, and washed into the dish by means of alcohol in a small wash bottle. The shell and foil are put to one side and subsequently weighed when dry. The composition in the dish is broken down quite fine with a flat-headed glass rod, and the alcohol evaporated on the water bath till the residue is moist, but not quite dry, 25 c.c. of water are then added, and the composition well stirred from the bottom. After the addition of 0.5 grm. of pure sodium, thiosulphate, the contents of the dish, is well stirred for two and a half minutes. One drop of methyl orange is then added, and the solution titrated with N/20 sulphuric acid, which has been standardised against weighings of 0.05-0.1 grm. fulminate to which 25 c.c. of water is added in a porcelain dish, then 0.5 grm. of thiosulphate, and after stirring for two and a half minutes, titrated with N/20 sulphuric acid. The small amount of antimony sulphide present does not interfere with the recognition of the end point. After titration, the solution is filtered through a small 5-1/2 cm. filter paper, which retains the antimony sulphide. The filter paper containing the Sb_2S_3 is well washed and then transferred to a large 6 by 1 test tube. Five c.c. of strong hydrochloric acid are added, and the contents of the tube boiled gently for a few seconds until the sulphide is dissolved and all the H_2S driven off or decomposed: 2-3 c.c. of a saturated solution of tartaric acid are added, and the contents of the tube washed into a 250 c.c. Erlenmeyer flask. The solution is then nearly neutralised with sodium carbonate, excess of bi-carbonate added, and after the addition of some starch solution titrated with N/20 iodine solution. This method for small quantities of stibnite is both quick and accurate, the error being about ±0.0003 grm. Sb_2S_3 at the outside.

The tendency of this method is to give slightly low figures for the fulminate, but since these are uniform within a negligible error, it does not affect the value of the results as a criterion of uniformity. The following test results were obtained by Dr Brownsdon:—

Fulminate Taken. Grm.	Fulminate Found. Grm.	Error. Grm.
0.0086	0.0083	-0.0003
0.0082	0.0081	-0.0001
0.0074	0.0071	-0.0003
0.0068	0.0066	-0.0002

Stibnite Taken. Grm.	Sb_2S_3, Found. Grm.	Error. Grm.
0.0085	0.0084	-0.0001
0.0098	0.0099	+0.0001
0.0160	0.0157	-0.0003
0.0099	0.0100	+0.0001

TABLE FOR CORRECTION OF VOLUMES OF GASES FOR TEMPERATURE, GIVING THE DIVISOR FOR THE FORMULA.

$V_1 = V \times B / (760 \times (1 + dt))$ ($d = 0.003665$) $1 + dt$ from 0° to 30° C.

t. °C.	760×(1+dt).	t. °C.	760×(1+dt).	t. °C.	760×(1+dt).
0.0	760.000	1.7	764.7352	3.4	769.4704
.1	760.2785	.8	765.0137	.5	769.7489
.2	760.5571	.9	765.2923	.6	770.0274
.3	760.8356	2.0	765.5708	.7	770.3060
.4	761.1142	.1	765.8493	.8	770.5845
.5	761.3927	.2	766.1279	.9	770.8631
.6	761.6712	.3	766.4064	4.0	771.1416
.7	761.9498	.4	766.6850	.1	771.4201
.8	762.2283	.5	766.9635	.2	771.6987
.9	762.5069	.6	767.2420	.3	771.9772
1.0	762.7854	.7	767.5206	.4	772.2558
.1	763.0639	.8	767.7991	.5	772.5343
.2	763.3425	.9	768.0777	.6	772.8128
.3	763.6210	3.0	768.3562	.7	773.0914
.4	763.8996	.1	768.6347	.8	773.3699
.5	764.1781	.2	768.9133	.9	

773.6485 .6 | 764.4566 | .3 | 769.1918 | 5.0 | 773.9270
___|_____|___|_____|___|_____

| | | | | t. | 760x(1+dt). | t. | 760x(1+dt). | t. | 760x(1+dt).
___|_____|___|_____|___|_____
| | | | | °C. | | °C. | | °C. | 5.1 | 774.2055 | .9 | 787.5755 | .7 |
800.9454 .2 | 774.4841 | 10.0 | 787.8540 | .8 | 801.2239 .3 | 774.7626
| .1 | 788.1325 | .9 | 801.5025 .4 | 775.0412 | .2 | 788.4111 | 15.0 |
801.7810 .5 | 775.3197 | .3 | 788.6896 | .1 | 802.0595 .6 | 775.5982 |
.4 | 788.9682 | .2 | 802.3381 .7 | 775.8768 | .5 | 789.2467 | .3 |
802.6166 .8 | 776.1553 | .6 | 789.5252 | .4 | 802.8952 .9 | 776.4339 |
.7 | 789.8038 | .5 | 803.1737 6.0 | 776.7124 | .8 | 790.0823 | .6 |
803.4522 .1 | 776.9909 | .9 | 790.3609 | .7 | 803.7308 .2 | 777.2695
| 11.0 | 790.6394 | .8 | 804.0093 .3 | 777.5480 | .1 | 790.9179 | .9 |
804.2879 .4 | 777.8266 | .2 | 791.1965 | 16.0 | 804.5664 .5 | 778.1051
| .3 | 791.4750 | .1 | 804.8449 .6 | 778.3836 | .4 | 791.7536 | .2 |
805.1235 .7 | 778.6622 | .5 | 792.0321 | .3 | 805.4020 .8 | 778.9407 |
.6 | 792.3106 | .4 | 805.6806 .9 | 779.2193 | .7 | 792.5892 | .5 |
805.9591 7.0 | 779.4978 | .8 | 792.8677 | .6 | 806.2376 .1 | 779.7763 |
.9 | 793.1463 | .7 | 806.5162 .2 | 780.0549 | 12.0 | 793.4248 | .8 |
806.7947 .3 | 780.3334 | .1 | 793.7033 | .9 | 807.0733 .4 | 780.6120 |
.2 | 793.9819 | 17.0 | 807.3518 .5 | 780.8905 | .3 | 794.2604 | .1 |
807.6303 .6 | 781.1690 | .4 | 794.5390 | .2 | 807.9089 .7 | 781.4476 |
.5 | 794.8175 | .3 | 808.1874 .8 | 781.7261 | .6 | 795.0960 | .4 |
808.4660 .9 | 782.0047 | .7 | 795.3746 | .5 | 808.7445 8.0 | 782.2832 |
.8 | 795.6531 | .6 | 809.0230 .1 | 782.5617 | .9 | 795.9317 | .7 |
809.3016 .2 | 782.8403 | 13.0 | 796.2102 | .8 | 809.5801 .3 | 783.1188
| .1 | 796.4887 | .9 | 809.8587 .4 | 783.3974 | .2 | 796.7673 | 18.0 |
810.1372 .5 | 783.6959 | .3 | 797.0458 | .1 | 810.4175 .6 | 783.9544 |
.4 | 797.3244 | .2 | 810.6943 .7 | 784.2330 | .5 | 797.6029 | .3 |
810.9728 .8 | 784.5115 | .6 | 797.8814 | .4 | 811.2514 .9 | 784.7901 |
.7 | 798.1600 | .5 | 811.5299 9.0 | 785.0686 | .8 | 798.4385 | .6 |
811.8084 .1 | 785.3471 | .9 | 798.7171 | .7 | 812.0870 .2 | 785.6257
| 14.0 | 798.9956 | .8 | 812.3655 .3 | 785.9042 | .1 | 799.2741 | .9 |
812.6441 .4 | 786.1828 | .2 | 799.5527 | 19.0 | 812.9226 .5 | 786.4613
| .3 | 799.8312 | .1 | 813.2011 .6 | 786.7398 | .4 | 800.1098 | .2 |
813.4797 .7 | 787.0184 | .5 | 800.3883 | .3 | 813.7582 .8 | 787.2969 |
.6 | 800.6668 | .4 | 814.0368
___|_____|___|_____|___|_____

t. °C.	760x(1+dt)	t. °C.	760x(1+dt)	t. °C.	760x(1+dt)
19.5	814.3153	23.0	824.0642	.5	833.8131
.6	814.5938	.1	824.3427	.6	834.0916
.7	814.8724	.2	824.6213	.7	834.3702
.8	815.1500	.3	824.8998	.8	834.6487
.9	815.4925	.4	825.1784	.9	834.9273
20.0	815.7080	.5	825.4569	27.0	835.2058
.1	815.9865	.6	825.7354	.1	835.4843
.2	816.2651	.7	826.0140	.2	835.7629
.3	816.5436	.8	826.2925	.3	836.0414
.4	816.8222	.9	826.5711	.4	836.3200
.5	817.1007	24.0	826.8496	.5	836.5985
.6	817.3792	.1	827.1281	.6	836.8770
.7	817.6578	.2	827.4067	.7	837.1556
.8	817.9363	.3	827.6852	.8	837.4341
.9	818.2149	.4	827.9638	.9	837.7127
21.0	818.4934	.5	828.2423	28.0	837.9912
.1	818.7719	.6	828.5208	.1	838.2697
.2	819.0505	.7	828.7994	.2	838.5483
.3	819.3290	.8	829.0779	.3	838.8268
.4	819.6076	.9	829.3565	.4	839.1054
.5	819.8861	25.0	829.6350	.5	839.3839
.6	820.1646	.1	829.9135	.6	839.6624
.7	820.4432	.2	830.1921	.7	839.9410
.8	820.7217	.3	830.4706	.8	840.2195
.9	821.0003	.4	830.7492	.9	840.4981
22.0	821.2788	.5	831.0277	29.0	840.7766
.1	821.5573	.6	831.3062	.1	841.0551
.2	821.8859	.7	831.5848	.2	841.3337
.3	822.1144	.8	831.8633	.3	841.6122
.4	822.3930	.9	832.1419	.4	841.8908
.5	822.6715	26.0	832.4204	.5	842.1693
.6	822.9500	.1	832.6989	.6	842.4478
.7	823.2286	.2	832.9775	.7	842.7264
.8	823.5071	.3	833.2560	.8	843.0049
.9	823.7857	.4	833.5346	.9	843.2835
		30.0			843.5620

CHAPTER VIII.

FIRING POINT OF EXPLOSIVES, HEAT TESTS, &c.

Horsley's Apparatus—Table of Firing points—The Government Heat-Test
Apparatus for Dynamites—Nitro-Glycerine, Nitro-Cotton, and Smokeless
Powders—Liquefaction and Exudation Tests—Page's Regulator for Heat-Test
Apparatus—Specific Gravities of Explosives—Table of Temperature of
Detonation, Sensitiveness, &c.

~The Firing Point of Explosives.~—The firing point of an explosive may be determined as follows:—A copper dish, about 3 inches deep, and 6 or more wide, and fitted with a lid, also of copper, is required. The lid contains several small holes, into each of which is soldered a thick copper tube about 5 mm. in diameter, and 3 inches long, with a rather larger one in the centre in which to place a thermometer. The dish is filled with Rose's metal, or paraffin, according to the probable temperature required. The firing point is then taken thus:—After putting a little piece of asbestos felt at the bottom of the centre tube, the thermometer is inserted, and a small quantity of the explosive to be tested is placed in the other holes; the lid is then placed on the dish containing the melted paraffin or metal, in such a way that the copper tubes dip below the surface of the liquid; the temperature of the bath is now raised until the explosive fires, and the temperature noted. The initial temperature should also be noted.

THE FIRING POINT OF VARIOUS EXPLOSIVES (by C. E. Munroe).
(Horsley's Apparatus used.)

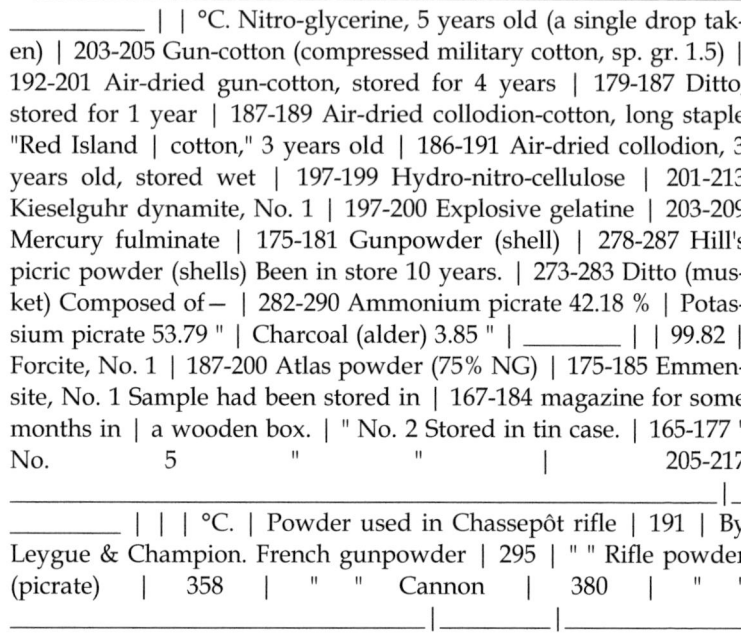

Horsley's apparatus consists of an iron stand with a ring support, holding a hemispherical iron vessel or bath in which solid paraffin is put. Above this is another movable support, from which a thermometer is suspended, and so adjusted that its bulb is immersed in the material contained in the iron vessel. A thin copper cartridge-case, 5/8 inch in diameter and 1-15/16 inch long, is suspended over the bath by means of a triangle, so that the end of the case is just 1 inch below the surface of the molten material. On beginning the experiment of determining the firing point of any explosive, the material in the bath is heated to just above the melting point; the thermometer is inserted in it, and a minute quantity of the explosive is placed in the bottom of the cartridge-case. The initial temperature is noted, and then the cartridge-case containing the explosive is inserted in the bath. The temperature is quickly raised until the contents of the cartridge-case flash off or explode, when the temperature is noted as the *firing point*.

[Illustration: FIG. 46.—HEAT TEST APPARATUS.]

Professor C.E. Munroe, of the U.S. Torpedo Station, has determined the firing point of several explosives by means of this apparatus.

~The Government Heat Test (Explosives Act, 1875): Apparatus required.~—A water bath, consisting of a spherical copper vessel *(a)*, Fig. 46, of about 8 inches diameter, and with an aperture of about 5 inches; the bath is filled with water to within a quarter of an inch of the edge. It has a loose cover of sheet copper about 6 inches in diameter *(b)* and rests on a tripod stand about 14 inches high *(c)*, which is covered with coarse wire gauze *(e)*, and is surrounded with a screen of thin sheet copper *(d)*. Within the latter is placed an argand burner *(f)* with glass chimney. The cover *(b)* has four holes arranged, as seen in Fig. II., No. 4 to contain a Page's[A] or Scheibler's regulator, No. 3 the thermometer, Nos. 1 and 2 the test tubes containing the explosive to be tested. Around the holes 1 and 2 on the under side of the cover are soldered three pieces of brass wire with points slightly converging (Fig. III.); these act as springs, and allow the test tubes to be easily placed in position and removed.

[Footnote A: See *Chem. Soc. Jour.*, 1876, i. 24. F.J.M. Page.]

~Test Tubes~, from 5-1/4 to 5-1/2 inches long, and of such a diameter that they will hold from 20 to 22 cubic centimetres of water when filled to a height of 5 inches; rather thick glass is preferable. Indiarubber stoppers, fitting the test tubes, and carrying an arrangement for holding the test papers, viz., a narrow glass tube passing through the centre of the stopper, and terminating in a platinum wire hook. A glass rod drawn out and the end turned up to form a hook is better.

~The Thermometer~ should have a range from 30° to 212° F., or from 1° to 100° C. A minute clock is useful.

~Test Paper.~—The test paper is prepared as follows:—45 grains (2.9 grms.) of white maize starch (corn flour), previously washed with cold water, are added to 8-1/2 oz. of water. The mixture is stirred, heated to boiling, and kept gently boiling for ten minutes; 15 grains (1 grm.) of pure potassium iodide (previously recrystallised from alcohol, absolutely necessary) are dissolved in 8-1/2 oz. of distilled water. The two solutions are thoroughly mixed and allowed to get cold. Strips or sheets of white English filter paper, previously washed with water and re-dried, are dipped into the solution thus prepared, and allowed to remain in it for not less than ten seconds; they are then allowed to drain and dry in a place free from laboratory fumes and dust. The upper and lower margins of the strips or sheets are cut off, and the paper is preserved in well- stoppered or corked bottles, and in the dark. The dimensions of the pieces of test paper used are about 4/10 inch by 8/10 inch (10 mm. by 20 mm.).[A]

[Footnote A: When the paper is freshly prepared, and as long as it remains in good condition, a drop of diluted acetic acid put on the paper with a glass rod produces no coloration. In process of time it will become brownish, when treated with the acid, especially if it has been exposed to sunlight. It is then not fit for use.]

In Germany zinc-iodide starch paper is used, which is considered to be more sensitive than potassium iodide.

~Standard Tint Paper.~ — A solution of caramel in water is made of such concentration that when diluted one hundred times (10 c.c. made up to 1 litre) the tint of this diluted solution equals the tint produced by the Nessler test in 100 c.c. water containing .000075 grm. of ammonia, or .00023505 grm. AmCl. With this caramel solution lines are drawn on strips of white filter paper (previously well washed with distilled water, to remove traces of bleaching matter, and dried) by means of a quill pen. When the marks thus produced are dry, the paper is cut into pieces of the same size as the test paper previously described, in such a way that each piece has a brown line across it near the middle of its length, and only such strips are preserved in which the brown line has a breadth varying from 1\2 mm. to 1 mm. (1/50 of an inch to 1/25 of an inch).

~Testing Dynamite, Blasting Gelatine, and Gelatine Dynamite.~ — Nitro- glycerine preparations, from which the nitro-glycerine can be extracted in the manner described below, must satisfy the following test, otherwise they will not be considered as manufactured with "thoroughly purified nitro-glycerine," viz., fifteen minutes at 160° F. (72° C.).

~Apparatus required.~ — A funnel 2 inches across (*d*), a cylindrical measure divided into grains (*e*), Fig. 47.

~Mode of Operation.~ — About 300 (19.4 grms.) to 400 grains (26 grms.) of dynamite (*b*), finely divided, are placed in the funnel, which is loosely plugged by freshly ignited asbestos (*a*). The surface is smoothed by means of a flat-headed glass rod or stopper, and some clean washed and dried kieselguhr (*c*) is spread over it to the depth of about 1/8 inch. Water is then poured on from a wash bottle, and when the first portion has been soaked up more is added; this is repeated until sufficient nitro- glycerine has collected in the graduated measure (*e*). If any water should have passed through, it must be removed from the nitro-glycerine by filter paper, or the nitro-glycerine may be filtered.

[Illustration: FIG. 47. — APPARATUS FOR SEPARATING THE NITRO-GLYCERINE FROM DYNAMITE.]

[Illustration: FIG. 48. — TEST TUBE ARRANGED FOR HEAT TEST.]

~Application of Test.~ — The thermometer is fixed so as to be inserted through the lid of the water bath into the water, which is maintained at 160° F. (72° C.), to a depth of 2-3/4 inches. Fifty grains (= 3.29 grms.) of nitro-glycerine to be tested are weighed into the test tube, in such a way as not to soil the sides of the tube (use a pipette). A test paper is fixed on the hook of the glass rod, so that when inserted into the tube it will be in a vertical position. A sufficient amount of a mixture of half distilled water and half glycerine, to moisten the upper half of the paper, is now applied to the upper edge of the test paper by means of a glass rod or camel's hair pencil; the cork carrying the rod and paper is fixed into the test tube, and the position of the paper adjusted so that its lower edge is about half way down the tube; the latter is then inserted through one of the holes in the cover to such a depth that the lower margin of the moistened part of the paper is about 5/8 inch above the surface cover. The test is complete when the faint brown line, which after a time makes its appearance at the line of boundary between the dry and moist part of the paper, equals in tint the brown line of the standard tint paper.

~Blasting Gelatine, Gelatine Dynamite, Gelignite, &c.~—Fifty grains (= 3.29 grms.) of blasting gelatine are intimately mixed with 100 grains (= 6.5 grms.) of French chalk. This is done by carefully working the two materials together with a wooden pestle in a wooden mortar. The mixture is then gradually introduced into the test tube, with the aid of gentle tapping upon the table between the introduction of successive portions of the mixture into the tube, so that when the tube contains all the mixture it shall be filled to the extent of 1-3/4 inch of its height. The test paper is then inserted as above described for nitro-glycerine. The sample tested must stand a temperature of 160° F. for a period of ten minutes before producing a discoloration of the test paper corresponding in tint to the standard paper.

N.B.—Non-gelatinised nitro-glycerine preparations, from which the nitro-glycerine cannot be expelled by water, are tested without any previous separation of the ingredients, the temperature being as above 160° F., and the time being seven minutes.

~Gun-Cotton, Schultze Gunpowder, E.C. Powder, &c.: A. Compressed Gun-Cotton.~—Sufficient material to serve for two or more tests is removed from the centre of the cartridge by gentle scraping, and if necessary, further reduced by rubbing between the fingers. The fine powder thus produced is spread out in a thin layer upon a paper tray 6 inches by 4-1/2 inches, which is then placed inside a water oven, kept as nearly as possible at 120° F. (49° C.). The wire gauze shelves of the oven should be about 3 inches apart. The sample is allowed to remain at rest for fifteen minutes in the oven, the door of which is left wide open. After the lapse of fifteen minutes the tray is removed and exposed to the air of the room for two hours, the sample being at some point within that time rubbed upon the tray with the hand, in order to reduce it to a fine and uniform state of division.

The heat test is performed as before, except that the temperature of the bath is kept at 170° F. (66° C.), and regulator set to maintain that temperature. Twenty grains (1.296 grm.) are used, placed in the test tube, gently pressed down until it occupies a space of as nearly as possible 1-5/10 inch in the test tube of dimensions previously specified. The fine cotton adhering to the sides of the tube can be removed by a clean cloth or silk handkerchief. The paper is moistened by touching the upper edge with a drop of the 50 per cent. glycerine solution, the tube inserted in the bath to a depth of 2-1/2 inches, measured from the cover, the regulator and thermometer being inserted to the same depth. The test paper is to be kept near the top of the test tube, but clear of the cork, until the tube has been immersed for about five minutes. A ring of moisture will about this time be deposited upon the sides of the test tube, a little above the cover of the bath. The glass rod must then be lowered until the lower margin of the moistened part of the paper is on a level with the bottom of the ring of moisture in the tube. The paper is now closely watched, The test is complete when a very faint brown coloration makes its appearance at the line of boundary between the dry and moist parts of the paper. It must stand the test for not less than ten minutes at 170° F. (The time is reckoned from the first insertion of the tube in the bath until the appearance of a discoloration of the test paper.)

~B. Schultze Powder, E.C. Powder, Collodion-Cotton, &c.~—The sample is dried in the oven as above for fifteen minutes, and exposed for two hours to the air. The test as above for compressed gun-cotton is then applied.

~C. Cordite~ must stand a temperature of 180° F. for fifteen minutes. The sample is prepared as follows:—Pieces half an inch long are cut from one end of every stick selected for the test: in the case of the thicker cordites, each piece so cut is further subdivided into about four portions. These cut pieces are then passed once through the mill, the first portion of material which passes through being rejected on account of the possible presence of foreign matter from the mill. The ground material is put on the top sieve of the nest of sieves, and sifted. That portion which has passed through the top sieve and been stopped by the second is taken for the test. If the mill is properly set, the greater portion of the ground material will be of the proper size. If the volatile matter in the explosive exceeds 0.5 per cent., the sifted material should be dried at a temperature not exceeding 140° F, until the proportion does not exceed 0.5 per cent. After each sample has been ground, the mill must be taken to pieces and carefully cleaned. The sieves used consist of a nest of two sieves with holes drilled in sheet copper. The holes in the top sieve have a diameter = 14 B.W.G., those in the second = 21 B.W.G.

If too hard for the mill, the cordite may be softened by exposure to the vapour of acetone,[A] or reduced, to the necessary degree of subdivision by means of a sharp moderately-coarse rasp. Should it have become too soft in the acetone vapour for the mill, it should be cut up into small pieces, which may be brought to any desired degree of hardness by simple exposure to air. Explosives which consist partly of gelatinised collodion-cotton, and partly of ungelatinised gun-cotton, are best reduced to powder by a rasp, or softened by exposure to mixed ether and alcohol vapour at a temperature of 40° F. to 100° F.

[Footnote A: Mr W. Cullen (*Jour. Soc. Chem. Ind.*, Jan. 31, 1901) says:— "Undoubtedly the advent of the horny smokeless powders of modern times has made it a little difficult to give the test the same scope as it had when first introduced." As a rule a simple explanation can be found for every apparently abnormal result, and in the accidental retention of a portion of the solvent used in the manufacture, will frequently be found an explanation of the trouble experienced.]

~Ballistite.~ — In the case of ballistite the treatment is the same, except that when it is in a very finely granulated condition it need not be cut up.

~Guttmann's Heat Test.~ — This test was proposed by Mr Oscar Guttmann in a paper read before the Society of Chemical Industry (vol. xvi., 1897), in the place of the potassium iodide starch paper used in the Abel test. The filter paper used is wetted with a solution of diphenylamine[A] in sulphuric acid. The solution is prepared as follows: — Take 0.100 grm. of diphenylamine crystals, put them in a wide-necked flask with a ground stopper, add 50 c.c. of dilute sulphuric acid (10 c.c. of concentrated sulphuric acid to 40 c.c. of water), and put the flask in a water bath at between 50° and 55° C. At this temperature the diphenylamine will melt, and at once dissolve in the sulphuric acid, when the flask should be taken out, well shaken, and allowed to cool. After cooling, add 50 c.c. of Price's double distilled glycerine, shake well, and keep the solution in a dark place. The test has to be applied in the following way:—The explosives that have to be tested are finely subdivided, gun-cotton, nitro-glycerine, dynamite, blasting gelatine, &c., in the same way as at present directed by the Home Office regulations. Smokeless powders are all to be ground in a bell-shaped coffee mill as finely as possible, and sifted as hitherto. 1.5 grm. of the explosive (from the second sieve in the case of smokeless powder) is to be weighed off and put into a test tube as hitherto used. Strips of well-washed filter paper, 25 mm. wide, are to be hung on a hooked glass rod as usual. A drop of the diphenylamine solution is taken up by means of a clean glass rod, and the upper corners of the filter paper are touched with it, so that when the two drops run together about a quarter of the filter paper is moist. This is then put into the test tube, and this again into the water bath, which has been heated to 70° C. The heat test reaction should not show in a shorter time than fifteen minutes. It will begin by the moist part of the paper acquiring a greenish yellow colour, and from this moment the paper should be carefully watched. After one or two minutes a dark blue mark will suddenly appear on the dividing line between the wet and dry part of the filter paper, and this is the point that should be taken.

[Footnote A: Dr G. Spica (*Rivista*, Aug. 1897) proposes to use hydrochloride of meta-phenylenediamine.]

~Exudation and Liquefaction Test for Blasting Gelatine, Gelatine Dynamite, &c.~—A cylinder of blasting gelatine, &c., is to be cut from the cartridge to be tested, the length of the cylinder to be equal to its diameter, and the ends being cut flat. The cylinder is to be placed on end on a flat surface without any wrapper, and secured by a pin passing vertically through its centre. In this condition the cylinder is to be exposed for 144 consecutive hours (six days and nights) to a temperature ranging from 85° to 90° F. (inclusive), and during such exposure the cylinder shall not diminish in height by more than one-fourth of its original height, and the upper cut surface shall retain its flatness and the sharpness of its edge.

~Exudation Test.~—There shall be no separation from the general mass of the blasting gelatine or gelatine dynamite of a substance of less consistency than the bulk of the remaining portion of the material under any conditions of storage, transport, or use, or when the material is subjected three times in succession to alternate freezing and thawing, or when subjected to the liquefaction test before described.

~Picric Acid.~ — The material shall contain not more than 0.3 part of mineral or non-combustible matter in 100 parts by weight of the material dried at 160° F. It should not contain more than a minute trace of lead. One hundred parts of the dry material shall not contain more than 0.3 part of total (free and combined) sulphuric acid, of which not more than 0.1 part shall be free sulphuric acid. Its melting point should be between 248° and 253° F.

~Ammonite, Bellite, Roburite, and Explosives of similar Composition.~ — These are required to stand the same heat test as compressed nitro-cellulose, gun-cotton, &c.

~Chlorate Mixtures.~ — The material must not be too sensitive, and must show no tendency to increase in sensitiveness in keeping. It must contain nothing liable to reduce the chlorate. Chlorides calculated as potassium chloride must not exceed 0.25 per cent. The material must contain no free acid, or substance liable to produce free acid. Explosives of this class containing nitro-compounds will be subject to the heat test.

~Page's Regulator.~ — The most convenient gas regulator to use in connection with the heat-test apparatus is the one invented by Prof. F.J.M. Page, B.Sc.[A] (Fig. 49). It is not affected by variations of the barometric pressure, and is simple and easy to fit up. It consists of a thermometer with an elongated glass bulb 5/8 inch diameter and 3 inches long. The stem of the thermometer is 5 inches long and 1/8 inch to 3/16 inch internal diameter. One and a half inch from the top of the stem is fused in at right angles a piece of glass tube, 1 inch long, of the same diameter as the stem, so as to form a T. A piece of glass tube (A), about 7/16 inch external diameter and 1-1/2 inch long, is fitted at one end with a short, sound cork (C, Fig. 50). Through the centre of this cork a hole is bored, so that the stem of the thermometer just fits into it. The other end of this glass tube is closed by a tightly fitting cork, preferably of indiarubber (I), which is pierced by a fine bradawl through the centre. Into the hole thus made is forced a piece of fine glass tube (B) 3 inches long, and small enough to fit loosely inside the stem of the thermometer.

[Footnote A: *Chemical Soc. Jour.*, 1876, i. 24.]

The thermometer is filled by pouring in mercury through a small funnel until the level of the mercury (when the thermometer is at the desired temperature) is about 1-1/2 inch below the T. The piece of glass tube A, closed at its upper extremity by the cork I, through which the fine glass tube B passes into the stem of the thermometer, is now filled by means of the perforated cork at its lower extremity on the stem of the thermometer. The gas supply tube is attached to the top of the tube A, the burner to the T, so that the gas passes in at the top, down the fine tube B, rises in the space between B and the inside wall of the stem of the thermometer, and escapes by the T. The regulator is set for any given temperature by pushing the cork C, and consequently the tubes A and B, which are firmly attached to it, up or down the stem of the thermometer, until the regulator just cuts off the gas at the desired temperature.

[Illustration: FIG. 49.—PAGE'S REGULATOR.]

[Illustration: FIG. 50.—PAGE'S GAS REGULATOR, SHOWING BYE-PASS AND CUT-OFF ARRANGEMENT.]

As soon as the temperature falls, the mercury contracts, and thus opens the end of the tube B. The gas is thus turned on, and the temperature rises until the regulator again cuts off the gas. In order to prevent the possible extinction of the flame by the regulator, the brass tube which carries the gas to the regulator is connected with the tube which brings the gas from the regulator to the burner by a small brass tap (Fig. 2). This tap forms an adjustable bye-pass, and thus a small flame can be kept burning, even though the regulator be completely shut off. It is obvious that the quantity of gas supplied through the bye-pass must always be less than that required to maintain the desired temperature. This regulator, placed in a beaker of water on a tripod, will maintain the temperature of the water during four or five hours within 0.2° C., and an air bath during six weeks within 0.5° C.

To sum up briefly the method of using the regulator:—Being filled with mercury to about 1\2 inch below the T, attach the gas supply as in diagram (Fig. 2), the brass tap being open, and the tube

B unclosed by the mercury. Allow the gas to completely expel the air in the apparatus. Push down the tube A so that the end of B is well under the surface of the mercury. Turn off the tap of the bye-pass until the smallest bead of flame is visible. Raise A and B, and allow the temperature to rise until the desired point is attained. Then push the tubes A and B slowly down until the flame is just shut off. The regulator will then keep the temperature at that point.

~Will's Test for Nitro-Cellulose.~ — The principle of Dr W. Will's test[A] may be briefly described as follows: — The regularity with which nitro- cellulose decomposes under conditions admitting of the removal of the products of decomposition immediately following their formation is a measure of its stability. As decomposing agent a sufficiently high temperature (135° C.) is employed, the explosive being kept in a constantly changing atmosphere of carbon dioxide, heated to the same temperature: the oxides of nitrogen which result are swept over red-hot copper, and are then reduced to nitrogen, and finally, the rates of evolution of nitrogen are measured and compared. Dr Will considers that the best definition and test of a stable nitro-cellulose is that it should give off at a high temperature equal quantities of nitrogen in equal times. For the purposes of manufacture, it is specially important that the material should be purified to its limit, i.e., the point at which further washing produces no further change in its speed of decomposition measured in the manner described.

[Footnote A: W. Will, *Mitt. a. d. Centrallstelle f. Wissench. Techn. Untersuchungen Nuo-Babelsberg Berlin*, 1902 [2], 5-24.]

The sample of gun-cotton (2.5 grms.) is packed into the decomposition tube 15 mm. wide and 10 cm. high, and heated by an oil bath to a constant temperature, the oxides so produced are forced over ignited copper, where they are reduced, and the nitrogen retained in the measuring tubes. Care must be taken that the acid decomposition products do not condense in any portion of the apparatus. The air in the whole apparatus is first displaced by a stream of carbon dioxide issuing from a carbon dioxide generator, or gas-holder, and passing through scrubbers, and this stream of gas is maintained throughout the whole of the experiment, the gas being absorbed at the end of the system by strong solution of caustic potash. To guard against the danger of explosions, which occasionally occur, the decomposition tube and oil bath are surrounded by a large casing with walls composed of iron plate and strong glass.

Dr Will's apparatus has been modified by Dr Robertson,[A] of the Royal Gunpowder Factory, Waltham Abbey. The form of the apparatus used by him is shown in Fig. 51.

~CO_2 Holders.~ — Although objection has been taken to the use of compressed CO_2 in steel cylinders on account of the alleged large and variable amount of air present, it has, nevertheless, been found possible to obtain this gas with as little as 0.02 per cent. of air. Frequent estimations of the air present in the CO_2 of a cylinder show that even with the commercial article, after the bulk of the CO_2 has been removed, the residual gas contains only a very small amount of air, which decreases in a gradual and perfectly regular manner. For example, one cylinder which gave 0.03 per cent. of air by volume, after three months' constant use gave 0.02 per cent. The advantage of using CO_2 from this source is obvious when compared with the difficulty of evolving a stream of gas of constant composition from a Kipps or Finkener apparatus. A micrometer screw, in addition to the main valve of the CO_2 cylinder, is useful for governing the rate of flow. A blank experiment should be made to ascertain the amount of air in the CO_2 and the correction made in the readings afterwards.

[Footnote A: *Jour. Soc. Chem. Ind.*, June 30, 1902, p. 819.]

[Illustration: Fig 51. — Will's Apparatus for Testing Nitro-cellulose]

~Measurement of Pressure and Rate of Flow.~ — Great attention is paid to the measurement of the rate of flow of gas, which is arrived at by counting with a stop-watch the number of bubbles of gas per minute in a small sulphuric acid wash bottle. A mercury manometer is introduced here, and is useful for detecting a leak in the apparatus. The rate of flow that gives the most satisfactory results is 1,000 c.c. per hour. If too rapid it does not become sufficiently preheated in the glass spiral, and if too slow there is a more rapid decomposition of the nitro-cellulose by the oxides of nitrogen which are not removed.

~Decomposition Tube.~ — This is of the form and dimensions given by Dr Will (15 mm. wide and 10 cm. high), the preheating worm being of the thinnest hydrometer stem tubing. The ground-in exit tube is kept in position by a small screw clamp with trunnion bearings.

~Bath.~ — To permit of two experiments being carried on simultaneously, the bath is adapted for two decomposition tubes, and is on the principle of Lothar Meyer's air bath, that is, the bath proper filled with a high- flashing hydrocarbon oil, and fitted with a lid perforated with two circular holes for the spiral tubes, is surrounded by an asbestos-covered envelope, in the interior of which circulate the products of combustion of numerous small gas jets. The stirrer, agitated by a water motor, or, better still, a hot-air engine, has a series of helical blades curved to give a thorough mixing to the oil. Great uniformity and constancy of temperature are thus obtained. The bath is fitted also with a temperature regulator and thermometer.

~Reduction Tube~ — This is of copper, and consists of two parts, the outer tube and an inner reaching to nearly the bottom of the former. Into the inner tube fits a spiral of reduced copper gauze, and into the annular space between the tubes is fitted a tightly packed reduced copper spiral. At the bottom the inlet tube dips into a layer of copper oxide asbestos, on the top of which is a layer of reduced copper asbestos. Through the indiarubber cork passes a glass tube, which leads the CO_2 and nitrogen out of the reduction tube. As the portion of the tube containing the spirals is heated to redness, water jackets are provided on both inner and outer tubes to protect the indiarubber cork.

~Nitrogen Measuring Apparatus.~ — The measuring tube with zigzag arrangement is used, having been found very economical in potash. It is most convenient to take readings by counterbalancing the column of potash solution and reading off the volume of gas at atmospheric pressure. For this purpose the tap immediately in front of the measuring tube is momentarily closed, this having been proved to be without ill effect on the progress of the test. In all experiments done by this test the air correction is subtracted from each reading, and the remainder brought to milligrams of nitrogen with the usual corrections. As objection has frequently been taken to the test on the ground of difficulty in interpreting the results obtained, Dr Robertson made a series of experiments for the purpose of standardising the test, and at the same time of arriving at the condition under which it could be applied in the most sensitive and efficient manner. A variety of nitro-celluloses having been tested, there were chosen as typical, of stable and unstable products, service gun-cotton on the one hand, and an experimental gun- cotton, Z, on the other. The first point brought out by these experiments was the striking uniformity of service gun-cotton, first in regard to the rectilinear nature of the curve of evolution of nitrogen, and secondly in regard to the small range within which a large number of results is included, 15 samples lying between 6.6 and 8.7 mgms. of nitrogen evolved in four hours. In the case of service gun-cotton, little difference in the rate of evolution of nitrogen evolved is obtained on altering the rate of passage of CO_2 gas through the wide range of 500 c.c. per hour to 2,500 c.c. per hour. With Z gun-cotton (see Fig. 52), however, the case is very different. Operating at a rate of 1,000 c.c. of CO_2 per hour, a curve of nitrogen evolution is obtained, which is bent and forms a good representation of the inherent instability of the material as proved to exist from other considerations. Operating at the rate of 1,500 c.c. per hour, as recommended by Dr Will, the evolution of nitrogen is represented by a straight line, steeper, however, than that of service gun-cotton. The rate of passage of CO_2 was therefore chosen at 1,000 c.c. per hour, or two-thirds of the rate of Dr Will, and this rate, besides possessing the advantage claimed of rendering diagnostic the manner of nitrogen evolution in Z gun-cotton, has in other cases been useful in bringing out relationships, which the higher rate would have entirely masked.

[Illustration: Fig. 52.—Dr. Robertson's results.]

[Illustration: Fig. 53.—Service Guncotton for Cordite made at a Private
Factory.]

Readings are taken thirty minutes from the time the nitro-cellulose is heated, and are taken at intervals of fifteen minutes for about four hours; fresh caustic potash is added every thirty minutes or so. It is convenient to plot the results in curves. The curves given in Fig. 53 are from gun-cotton manufacturers in England at a private factory. The rate of evolution of nitrogen is as follows: —

In 1 hour. In 2 hours. In 3 hours. In 4 hours. N. N. N. N. in milligrammes. 1.25 2.55 4.5 5.75 1.5 3.25 5.25 6.75 These results are very satisfactory, the gun-cotton was of a very good quality. Several hours are necessary to remove all the air from the apparatus. Dr Will stated fifteen minutes in his original paper, but this has not been found sufficient. It has not been satisfactorily proved that Will's test can be applied to gelatinised nitro-cellulose powders. It is convenient to plot the results in curves; the nitrogen is generally given in cubic centimetres or in milligrammes, and readings taken every fifteen minutes. The steepness of the curve is a measure of the stability of the nitro-cellulose which is being examined. The steeper the curve the more nitrogen is evolved per unit of time, and the less stable the nitro- cellulose. In the case of unstable nitro-celluloses heated under the conditions described, the separation of nitrogen is much greater at first than at a later period. If the nitro-cellulose be very unstable, explosions are produced. If the separation of nitrogen is uniform during the prolonged heating, then the nitro-cellulose may be regarded as "normal." If it be desired to determine the absolute amount of nitrogen separated from a nitro-cellulose, the following conditions must be observed: — (1.) Accurate weighing of the nitro-cellulose; (2.) Determination of the amount of air in the CO_2, and deduction of this from the volume of gas obtained; (3.) Reduction of the volume of the gas to the volume at 0° C. and 760 mm. pressure.[A]

[Footnote A: See also *Jour. Soc. Chem. Ind.*, Dec. 1902, pages 1545-1555, on the "Stability of Nitro-cellulose" and "Examination of Nitro-cellulose," Dr Will.]

~Bergrnann and Junk~[A] describe a test for nitro-cellulose that has been in use in the Prussian testing station for some years. The apparatus consists of a closed copper bath provided with a condenser and 10 countersunk tubes of 20 cm. length. By boiling amyl-alcohol in the bath, the tubes can be kept at a constant temperature of 132° C. The explosive to be tested is placed in a glass tube 35 cm.

long and 2 cm. wide, having a ground neck into which an absorption bulb is fitted. The whole apparatus is surrounded by a shield, in case of explosion. In carrying out the test, 2 grms. of the explosive are placed in the glass tube and well pressed down. The absorption bulb is half filled with water, and fitted into the ground neck of the glass tube, which is then placed in one of the tubes in the bath previously brought to the boiling point (132° C.). The evolved oxides of nitrogen are absorbed in the water in the bulb, and at the end of two hours the tubes are removed from the bath, and on cooling, the water from the bulb flows back and wets the explosive. The contents of the tube are filtered and washed, the filtrate is oxidised with permanganate, and the nitrogen determined as nitric oxide by the Schultze-Tieman method. The authors conclude that a stable guncotton does not evolve more than 2.5 c.c. of nitric oxide per grm. on being heated to 132° C. for two hours, and a stable collodion-cotton not more than 2 c.c. under the same conditions. The percentage of moisture in the sample to be tested should be kept as low as possible. A sample of nitro-cellulose containing 1.97% of moisture gave an evolution of 2.6 c.c. per grm., while the same sample with 3.4% moisture gave an evolution of over 50 c.c. per grm. Sodium carbonate added to an unstable nitro-cellulose diminishes the rate of decomposition, but if sodium carbonate be intimately mixed with a stable nitro-cellulose the rate of decomposition will be increased. Calcium carbonate and mercury chloride have no influence. If an unstable nitro- cellulose be extracted with alcohol a stable compound is produced. The percentage solubility of a nitro-cellulose in ether-alcohol rises on heating to 132° C. A sample which before heating had a solubility of 4.7% had its solubility increased to 82.5% after six hours' heating.

[Footnote A: *Jour. Soc. Chem. Ind.*, xxiii., Oct. 15, 1904, p. 953.]

Mr A.P. Sy (*Jour. Amer. Chem. Soc.*, 1903) describes a new stability test for nitro-cellulose which he terms "The Elastic Limit of Powder Resistance to Heat." The test consists in heating the powder on a watch glass in an oven to a temperature of 115° C., after eight hours the watch glass and powder are weighed and the process repeated daily for six days or less. He claims that the powder is tested in its natural state, all the products of decomposition are taken into account, whilst in the old tests only the acid products are shown, and in the Will test only nitrogen, that it affords an indication of the effect of small quantities of added substances or foreign matters on the stability and that it is simple, and not subject to the variations of the old tests.

Obermüller (*Jour. Soc. Chem. Ind.*, April 15, 1905) considers Bergmann and Junk's test is too complicated and occupies too much time; he proposes to heat gun-cotton to 140° C. *in vacuo,* and to measure continuously by means of a mercury manometer the pressure exerted by the evolved gases, the latter being maintained at constant volume; the rate at which the pressure increases is a measure of the rate of decomposition of the nitro- cellulose.

SPECIFIC GRAVITIES OF EXPLOSIVES, &C.

Nitro-glycerine 1.6
Gun-cotton (dry) 1.06
  " (25 per cent. water) 1.32
Dynamite No. 1 1.62
Blasting gelatine 1.54
Gelatine dynamite 1.55
Ballistite 1.6
Forcite 1.51
Tonite 1.28
Roburite 1.40
Bellite 1.2-1.4
Carbo-dynamite 1.5
Turpin's cast picric acid 1.6
Nitro-mannite 1.6
Nitro-starch 1.5
Emmensite 1.8
Mono-nitro-benzene 1.2
Meta-di-nitro-benzene 1.575 at 18° C.
Ortho-di-nitro-benzene 1.590 "
Para-di-nitro-benzene 1.625 "
British gunpowder, E.X.E. 1.80
  " " S.B.C. 1.85
Cannonite (powder) 1.60
Celluloid 1.35
Cellulose 1.45

Ammonium nitrate 1.707
Mercury fulminate 4.42

TABLE OF THE TEMPERATURE OF DETONATION.

Blasting gelatine 3220°
Nitro-glycerine 3170°
Dynamite 2940°
Gun-cotton 2650°
Tonite 2648°
Picric acid 2620°
Roburite 2100°
Ammonia nitrate 1130°

RELATIVE SENSITIVENESS TO DETONATION (by Professor C.E. Munroe, U.S. Naval Torpedo Station).

Maximum Distance at which Detonation occurred. CM.	
10	Nitro-glycerine 86.5 nitro-cotton 9.5, camphor 4 per cent.
20	Explosive gelatine
	NH_4NO_3 5 parts, (camphorated) $C_6H_4(NO_3)_2$ 1 part.
25	Judson powder, R.R.P.
30	*Emmensite (No. 259)*
32	*Rack-a-rock*
	$KClO_3$ 79 parts, $C_6H_5(NO)_2$ 21 parts. Bellite
50	Forcite No. 1
61	Kieselguhr dynamite No. 1
64	75 per cent. nitro-gycerine. Atlas powder No. 1
74	

CHAPTER IX.

DETERMINATION OF THE RELATIVE STRENGTH OF EXPLOSIVES.

Effectiveness of an Explosive—High and Low Explosives—Theoretical
Efficiency—MM. Roux and Sarrau's Results—Abel and Noble's—Nobel's
Ballistic Test—The Mortar, Pressure, or Crusher Gauge—Lead Cylinders—
The Foot-Pounds Machine—Noble's Pressure Gauge—Lieutenant Walke's
Results—Calculation of Pressure Developed by Dynamite and Gun-Cotton—
Macnab's and Ristori's Results of Heat Developed by the Explosion of
Various Explosives—Composition of some of the Explosives in Common Use
for Blasting, &c.

~The Determination of the Relative Strength of Explosives.~—Explosives may be roughly divided into two divisions, viz., those which when exploded produce a shattering force, and those which produce a propulsive force. Explosives of the first class are generally known as the high explosives, and consist for the most part of nitro compounds, or mixtures of nitro compounds with other substances. Any explosive whose detonation is very rapid is a high explosive, but the term has chiefly been applied to the nitro-explosives.

The effectiveness of an explosive depends upon the volume and temperature of the gases formed, and upon the rapidity of the explosion. In the high explosives the chemical transformation is very rapid, hence they exert a crushing of shattering effect. Gunpowder, on the other hand, is a low explosive, and produces a propelling or heaving effect.

The maximum work that an explosive is capable of producing is proportionate to the amount of heat disengaged during its chemical transformation. This may be expressed in kilogrammetres by the formula 425Q, where Q is the number of units of heat evolved. The theoretical efficiency of an explosive cannot, however, be expected in practice for many reasons.

In the case of blasting rock, for instance:[A] — 1. Incomplete combustion of the explosive. 2. Compression and chemical changes induced in the surrounding material operated on. 3. Energy expended in the cracking and heating of the material which is not displaced. 4. The escape of gas through the blast-hole, and the fissures caused by the explosion. The proportion of useful work has been estimated to be from 14 to 33 per cent. of the theoretical maximum potential.

[Footnote A: C.N. Hake, Government Inspector of Explosives, Victoria, *Jour. Soc. Chem. Ind.*, 1889.]

For the purposes of comparison, manufacturers generally rely more upon the practical than the theoretical efficiency of an explosive. These, however, stand in the same relation to one another, as the following table of Messrs Roux and Sarrau will show: —

MECHANICAL EQUIVALENT OF EXPLOSIVES.

	Theoretical Work in Kilos.	Relative Value.
Blasting powder (62 per cent. KNO_3)	242,335	1.0
Dynamite (75 per cent. nitro-glycerine)	548,250	2.26
Blasting gelatine (92 per cent. nitro-glycerine)	766,813	3.16
Nitro-glycerine	794,563	3.28

Experiments made in lead cylinders give —
  Dynamite 1.0
  Blasting gelatine 1.4
  Nitro-glycerine 1.4

Sir Frederick Abel and Captain W.H. Noble, R.A., have shown that the maximum pressure exerted by gunpowder is equal to 486 foot-tons per lb. of powder, or that when 1 kilo, of the powder gases occupy the volume of 1 litre, the pressure is equal to 6,400 atmospheres; and Berthelot has calculated that every gramme of nitro-glycerine exploded gives 1,320 units of heat. MM. Roux and Sarrau, of the Depôt Centrales des Poudres, Paris, by means of calorimetric determinations, have shown that the following units of heat are produced by the detonation of —

Nitro-glycerine 1,784 heat units.
Gun-cotton 1,123 "
Potassic picrate 840 "

which, multiplied by the mechanical equivalent per unit, gives —

Nitro-glycerine 778 metre tons per kilogramme.
Gun-cotton 489 " "
Picrate of potash 366 " "

~Nobel's Ballistic Test.~ — Alfred Nobel was the first to make use of the mortar test to measure the (ballistic) power of explosives. The use of the mortar for measuring the relative power of explosives does not give very accurate results, but at the same time the information obtained is of considerable value from a practical point of view. The mortar consists of a solid cylinder of cast iron, one end of which has been bored to a depth of 9 inches, the diameter of the bore being 4 inches. At the bottom of the bore-hole is a steel disc 3 inches thick, in which another hole has been bored 3 inches by 2 inches. The mortar (Fig. 54) itself is fitted with trunnions, and firmly fixed in a very solid wooden carriage, which is securely bolted down to the ground. The shot used should weigh 28 lbs., and be turned accurately to fit the bore of the mortar. Down its centre is a hole through which the fuse is put.

The following is the method of making an experiment: — A piece of hard wood is turned in the lathe to exactly fit the hole in the steel disc at the bottom of the bore. This wooden cylinder itself contains a small cavity into which the explosive is put. Ten grms. is a very convenient quantity. Before placing in the mortar, a hole may be made in the explosive by means of a piece of glass rod of such a size that the detonator to be used will just fit into it. After placing the wooden cylinder containing the explosive in the cavity at the bottom of the bore, the shot, slightly oiled, is allowed to fall gently down on to it. A piece of fuse about a foot long, and fitted with a detonator, is now pushed through the hole in the centre of the shot until the detonator is embedded in the explosive. The fuse is now lighted, and the distance to which the shot is thrown is carefully measured. The range should be marked out with pegs into yards and fractions of yards, especially at the end opposite to the mortar. The mortar should be inclined at an angle of 45°. In experimenting with this apparatus, the force and direction of the wind will be found to have considerable influence.

[Illustration: FIG. 54. — MORTAR FOR MEASURING THE BALLISTIC POWER OF
EXPLOSIVES. A, Shot; B, Steel Disc; C, Section of Mortar (Cast Iron); D, Wooden Plug holding Explosive (E); F, Fuse.]

Mr T. Johnson made some ballistic tests. He used a steel mortar and a shot weighing 29 Ibs., and he adopted the plan of measuring the distance to which a given charge, 5 grms., would throw the shot. He obtained the following results: —

Range in Feet.

Blasting gelatine (90 per cent. nitro-glycerine and nitro-cellulose) 392
Ammonite (60 per cent. $Am(NO_3)$ and 10 per cent. nitro-naphthalene) 310
Gelignite (60 per cent. nitro-gelatine and gun-cotton) 306
Roburite ($AmNO_3$ and chloro-nitro-benzol) 294
No. 1 dynamite (75 per cent. nitro-gelatine) 264
Stonite (68 per cent. nitro-gelatine and 32 per cent. wood-meal) 253
Gun-cotton 234
Tonite (gun-cotton and nitrates) 223
Carbonite (25 per cent. nitro-gelatine, 40 per cent. wood-meal, and 30 per cent. nitrates) 198
Securite (KNO_3 and nitro-benzol) 183
Gunpowder 143

~Calculation of the Volume of Gas Evolved in an Explosive Reaction.~ — The volume of gas evolved in an explosive reaction may be calculated, but only when they are simple and stable products, such calculations being made at 0° and 760 mm. Let it be required, for example, to determine the volume of gas evolved by 1 gram-molecule of nitro-glycerine. The explosive reaction of nitro-glycerine may be represented by the equation.

$$C_3H_5O_3(NO_2)_3 = 3CO_2 + 2\text{-}1/2\,H_2O + 1\text{-}1/2\,N_2 + 1/4\,O_2$$

By weight 227 = 132 + 45 + 42 + 8
By volume 2 = 3 + 2-1/2 + 1-1/2 + 1/4

The weights of the several products of the above reactions are calculated by multiplying their specific gravities by the weight of 1 litre of hydrogen at 0° C. and 760 mm. (0.0896 grm). Thus,

One litre of CO_2 = 22 × .0896 = 1.9712 grm.
" H_2O = 9 × " = 0.8064 "
" N_2 = 14 × " = 1.2544 "
" O_2 = 16 × " = 1.4336 "

The volume of permanent gases at 0° and 760 mm. is constant, and assuming the gramme as the unit of mass, is found to be 22.32 litres. Thus: —

Volume of 44 of CO_2, at 0° and 760 mm. = 44/1.9712 = 22.32 litres. 18 " H_2O " " = 18/0.8044 = 22.32 " 28 " N_2 " " = 28/1.2544 = 22.32 " 32 " O_2 " " = 32/1.4366 = 22.32 "

Therefore

132 grms. of CO_2 at 0° C and 760 mm. = 22.32 × 3 = 66.96 litres.
 45 " H_2O " " = 22.32 × 2-1/2 = 55.80 "
 42 " N_2 " " = 22.32 × 1-1/2 = 33.48 "
  8 " O_2 " " = 22.32 × 1/4 = 5.58 "

161.82 " Therefore 1 gram-molecule or 227 grms. of nitro-glycerine when exploded, produces 161.82 litres of gas at 0° C and 760 mm.

To determine the volume of gas at the temperature of explosion, we simply apply the law of Charles.[A] Thus—

$V : V' :: T : T'$ or $V' = VT'/T$

in which V represents the original volume.

V' " new volume.

T " original temperature on the absolute scale.

T' " new temperature of the same scale

In the present case $T' = 6001°$.

Therefore substituting, we have

$V' = 161.82 \times 6001 / 273 = 3557$ litres

or at the temperature of explosion 1 gram-molecule of nitroglycerine produces 3,557 litres of permanent gas.

[Footnote A: According to the law of Charles, the volume of any gas varies directly as its temperature on the absolute scale, provided the pressure remains constant. Knowing the temperature on the centigrade scale, the corresponding temperature on the absolute scale is obtained by adding 273 to the degrees centigrade.]

~Pressure or Crusher Gauge.~—There are many forms of this instrument. As long ago as 1792 Count Rumford used a pressure gauge. The so-called crusher gauge was, however, first used by Captain Sir Andrew Noble in his researches on powder. Other forms are the Rodman[A] punch Uchatius Eprouvette, and the crusher gauge of the English Commission on Explosives. They are all based either upon the size of an indent made upon a copper disc by a steel punch fitted to a piston, acted upon by the gases of the explosive, or upon the crushing or flattening of copper or lead cylinders.

[Footnote A: Invented by General Rodman, United States Engineers.]

[Illustration: FIG. 55.—PRESSURE GAUGE.]

Berthelot uses a cylinder of copper, as also did the English Commission, but in the simpler form of apparatus mostly used by manufacturers lead cylinders are used. This form of apparatus (Fig. 55) consists of a base of iron to which four uprights *a* are fixed, set round the circumference of a 4-inch circle; the lead plug rests upon the steel base let into the solid iron block. A ring c holds the uprights d together at the top. The piston b, which rests upon the lead plug, is a cylinder of tempered steel 4 inches in diameter and 5 inches in length; it is turned away at the sides to lighten it as much as possible. It should move freely between the uprights d. In the top of this cylinder is a cavity to hold the charge of explosive. The weight of this piston is 12-1/4 lbs. The shot e is of tempered steel, and 4 inches in diameter and 10 inches in length, and weighs 34-1/2 lbs. It is bored through its axis to receive a capped fuse.

The instrument is used in the following manner: — A plug of lead 1 inch long and 1 inch in diameter, and of a cylindrical form, is placed upon the steel plate between the uprights a, the piston placed upon it, the carefully weighed explosive placed in the cavity, and the shot lowered gently upon the piston. A piece of fuse, with a detonator fixed at one end, is then pushed through the hole in the shot until it reaches the explosive contained in the cavity in the piston. The fuse is lighted. When the charge is exploded, the shot is thrown out, and the lead cylinder is more or less compressed. The lead plugs must be of a uniform density and homogeneous structure, and should be cut from lead rods that have been drawn, and not cast separately from small masses of metal.

[Illustration: FIG. 56.—b, STEEL PUNCH; c, LEAD CYLINDER FOR USE WITH
PRESSURE GAUGE.]

The strength of the explosive is proportional to the work performed in reducing the height of the lead (or copper) plug, and to get an expression for the work done it is necessary to find the number of foot-pounds (or kilogrammetres) required to produce the different amounts of compression. This is done by submitting exactly similar cylinders of lead to a crushing under weights acting without initial velocity, and measuring the reduced heights of the cylinders; from these results a table is constructed establishing empirical relations between the reduced heights and the corresponding weights; the cylinders are measured both before and after insertion in the pressure gauge by means of an instrument known as the micrometer calipers (Fig. 57).[A]

[Footnote A: An instrument called a "Foot-pounds Machine" has been invented by Lieut. Quinan, U.S. Army. It consists of three boards, connected so as to form a slide 16 feet high, in which a weight (the shot of the pressure gauge) can fall freely. One of the boards is graduated into feet and half feet. The horizontal board at the bottom, upon which the others are nailed, rests upon a heavy post set deep in the ground, upon which is placed the piston of the gauge, which in this case serves as an anvil on which to place the lead cylinders. The shot is raised by means of a pulley, fixed at the top of the structure, to any desired height, and let go by releasing the clutch that holds it. The difference between the original length and the reduced length gives the compression caused by the blow of the shot in falling, and gives the value in foot-pounds required to produce the different amounts of compression. (Vide *Jour. U.S. Naval Inst.*, 1892.)]

[Illustration: FIG. 57.—MICROMETER CALIPERS FOR MEASURING DIAMETER OF
LEAD CYLINDERS.]

~The Use of Lead Cylinders.~—The method of using lead cylinders to test the strength of an explosive is a very simple affair, and is conducted as follows:—A solid cast lead cylinder, of any convenient size, is bored down the centre for some inches, generally until the bore-hole reaches to about the centre of the block. The volume of this hole is then accurately measured by pouring water into it from a graduated measure, and its capacity in cubic centimetres noted. The bore-hole is then emptied and dried, and a weighed quantity (say 10 grms.) of the explosive pressed well down to the bottom of the hole. A hole is then made in the explosive (if dynamite) with a piece of clean and rounded glass rod, large enough to take the detonator. A piece of fuse, fitted with a detonator, is then inserted into the explosive and lighted. After the explosion a large pear-shaped cavity will be found to have been formed, the volume of which is then measured in the same way as before.

The results thus obtained are only relative, but are of considerable value for comparing dynamites among themselves (or gun-cottons). Experiments in lead cylinders gave the relative values for nitroglycerine 1.4, blasting gelatine 1.4, and dynamite 1.0. (Fig. 58 shows sections of lead cylinders before and after use.)

[Illustration: FIG. 58. — LEAD CYLINDERS BEFORE AND AFTER USE.]

Standard regulations for the preparation of lead cylinders may be found in the *Chem. Zeit.*, 1903, 27 [74], 898. They were drawn up by the Fifth International Congress of App. Chem., Berlin. The cylinder of lead should be 200 mm. in height and 200 mm. in diameter. In its axis is a bore-hole, 125 mm. deep and 25 mm. in diameter. The lead used must be pure and soft, and the cylinder used in a series of tests must be cast from the same melt. The temperature of the cylinders should be 15° to 20° throughout. Ten grms. of explosive should be used and wrapped in tin-foil. A detonator with a charge of 2 grms., to be fired electrically, is placed in the midst of the explosive. The cartridge is placed in the bore-hole, and gently pressed against the bottom, the firing wires being kept in central position. The bore-hole is then filled with dry quartz sand, which must pass through a sieve of 144 meshes to the sq. cm., the wires being .35 mm. diameter. The sand is filled in evenly, any excess being levelled off. The charge thus prepared is then fired electrically. The lead cylinder is then inverted, and any residues removed with a brush. The number of c.c. of water required to fill the cavity, in excess of the original volume of the bore-hole, is a measure of the strength of the explosive. The results are only comparable if made with the same class of explosive. A result is to be the mean of at least three experiments. The accuracy of the method depends on (*a*) the uniform temperature of the lead cylinder (15° to 20° C. 7); (*b*) on the uniformity of the quartz sand; (*c*) on the uniformity of the measurements.

[Illustration: FIG. 59. — NOBLE'S PRESSURE GAUGE.]

~Noble's Pressure Gauge.~ — The original explosive vessels used by Captain Sir A. Noble in his first experiments were practically exactly similar to those that he now employs, which consists of a steel barrel A (Fig. 59), open at both ends, which are closed by carefully fitted screw plugs, furnished with steel gas checks to prevent any escape past the screw. The action of the gas checks is exactly the same as the leathers used in hydraulic presses. The pressure of the gas acting on both sides of the annular space presses these sides firmly against the cylinder and against the plug, and so effectually prevents any escape. In the firing plug F is a conical hole closed by a cone fitting with great exactness, which, when the vessel is prepared for firing, is covered with fine tissue paper to act as an insulator. The two firing wires GG, one in the insulated cone, the other in the firing plug, are connected by a very fine platinum wire passing through a glass tube filled with meal powder. The wire becomes red-hot when connection is made with a Leclanché battery, and the charge which has previously been inserted into the vessel is fired. The crusher plug is fitted with a crusher gauge H for determining the pressure of the gases at the moment of explosion, and in addition there is frequently a second crusher gauge apparatus screwed into the cylinder. When it is desired to allow the gases to escape for examination, the screw J is slightly withdrawn. The gases then pass into the passage I, and can be led to suitable apparatus in which their volume can be measured, or in which they can be sealed for subsequent chemical analysis.

The greatest care must be exercised in carrying out experiments with this apparatus; it is particularly necessary to be sure that all the joints are perfectly tight before exploding the charge. Should this not be the case, the gases upon their generation will cut their way out, or completely blow out the part improperly secured, in either case destroying the apparatus. The effect produced upon the apparatus when the gas has escaped by cutting a passage for itself is very curious. The surface of the metal where the escape occurred presents the appearance of having been washed away in a state of fusion by the rush of the highly heated products.

~The Pressure Gauge.~ — The pressure is found by the use of a little instrument known as the pressure gauge which consists of a small chamber formed of steel, inside of which is a copper cylinder, and the entrance being closed by a screw gland, in which a piston, having a definite sectional area, works. There is a gas check E (Fig. 60) placed in the gland, and over the piston, which prevents the admission of gas to the chamber. When it is desired to find the pressure in the chamber of a gun, one or more of these crushers are made up with or inserted at the extreme rear end of the cartridge, in order to avoid their being blown out of the gun when fired. This, however, often takes place, in which case the gauges are usually found a few yards in front of the muzzle. The copper cylinders which register the pressure are made 0.5 inch long from specially selected copper, the diameters being regulated to give a sectional area of either 1/12 or 1/24 square inch.

[Illustration: FIG. 60. — CRUSHER GAUGE. *E*, GAS CHECK.]

Hollow copper cylinders are manufactured with reduced sectional areas for measuring very small pressures. It has been found that these copper cylinders are compressed to definite lengths for certain pressures with remarkable uniformity. Thus a copper cylinder having a sectional area of 1/12 square inch, and originally 1/2 inch long, is crushed to a length of 0.42 inch by a pressure of 10 tons per square inch. By subsequently applying a pressure of 12 tons per square inch the cylinder is reduced to a length of 0.393 inch. Before using the cylinders, whether for experimenting with closed vessels or with guns, it is advisable to first crush them by a pressure a little under that expected in the experiment. Captain Sir A. Noble used in his experiments a modification of Rodman's gauge. (Ordnance Dept., U.S.A., 1861.)

~By Calculation.~ — To calculate the pressure developed by the explosion of dynamite in a bore-hole 3 centimetres in diameter, charged with 1 kilogramme of 75 per cent. dynamite, Messrs Vieille and Sarrau employ the following formula: —

$P = V_o(1 + Q/273.c)/(V - v)$.

Where V_o = the volume (reduced to 0° and 760 mm.) of the gases produced by a unit of weight of the explosive; Q the number of calories disengaged by a unit of weight of the explosive; *c* equals the specific heat at constant volume of the gases; V the volume in cubic centimetres of a unit of weight of the explosive; *v* the volume occupied by the inert materials of the explosive. The volume of gas produced by the explosion of 1 kilogramme of nitro-glycerine (at 0° and 760 mm.) is 467 litres.

V_o will therefore equal $0.75 \times 467 = 350.25$.

The specific heat c is, according to Sarrau, .220 (c); and according to Bunsen, 1 kilogramme of dynamite No. 1 disengages 1,290 (Q) calories. The density of dynamite is equal to 1.5, therefore

V = 1/1.5 = .666.

If we take the volume of the kieselguhr as .1, we find from above formula that

P = 350(1 + 1290/(273 x .222))/(.600 - .1) = 13,900 atmospheres,

which is equal to 14,317 kilogrammes per square centimetre. The pressure developed by 1 kilogramme of pure nitro-glycerine equals 18,533 atmospheres, equals 19,151 kilogrammes. Applying this formula to gun-cotton, and taking after Berthelot, Q = 1075, and after Vieille and Sarrau, V_{o} = 671 litres, and c as .2314, and the density of the nitro-cellulose as 1.5, we have (V = O)

P = 671(1 + 1075/(273 x .2314))/.666 = 18,135 atmospheres.

To convert this into pressure of kilogrammes per square centimetre, it is necessary to multiply it by the weight of a column of mercury 0.760 m. high, and 1 square centimetre in section, which is equal to increasing it by 1/30. It thus becomes

P^{k} = (1 + 1/30).

P^{k} = 18,135 x 1.033 = 18,733 kilogrammes.

The following tables, taken from Messrs William Macnab's and E. Ristori's paper (*Proc. Roy. Soc.*, 56, 8-19), "Researches on Modern Explosives," are very interesting. They record the results of a large number of experiments made to determine the amount of heat evolved, and the quantity and composition of the gases produced when certain explosives and various smokeless powders were fired in a closed vessel from which the air had been previously exhausted. The explosions were carried out in a "calorimetric bomb" of Berthelot's pattern.[A]

[Footnote A: For description of "bomb," see "Explosives and their Power,"
Berthelot, trans. by Hake and Macnab, p. 150. (Murray.)]

Table Showing Quantity of Heat and Volume and Analysis of Gas Developed per Gramme with Different Sporting and Military Smokeless Powders Now In Use

Name of Explosive.	Calories per grm.	Permanent Gases. cc/grm	Aqueous Vapour. cc/grm	Total Volume of Gas at 0° and 760 mm. cc/grm
E.C. powder, English	800	420	154	574
S.S. powder	799	584	150	734
Troisdorf, German	943	700	195	895
Rifleite, English	864	766	159	925
B.N., French	833	738	168	906
Cordite, English	1253	647	235	882
Ballistite, German	1291	591	231	822
Ballistite, Italian and Spanish	1317	58l	245	826

The figures in column headed "Co-efficient of Potential Energy" serve as a measure of comparison of the power of the explosives, and are the products of the number of calories by the volume of gas, the last three figures being suppressed in order to simplify the results.

The amounts of water found were calculated for comparison as volumes of H_2O gas at 0° and 760 mm.

E.C. powder consists principally of nitro-cellulose mixed with barium nitrate and a small proportion of camphor.

S.S. of nitro-lignine mixed with barium nitrate and nitro-benzene.

Troisdorf powder is gelatinised nitro-cellulose; rifleite gelatinised nitro-cellulose and nitro-benzene.

Cordite contains 58 per cent. nitro-glycerine, 37 per cent. gun-cotton, and 5 per cent. vaseline.

Ballistite (Italian) consists of equal parts nitro-cellulose and nitro-glycerine, and 1/2 per cent. of aniline. The German contains a higher percentage of nitro-cellulose.

TABLE SHOWING THE HEAT DEVELOPED BY EXPLOSIVES CONTAINING NITRO-GLYCERINE AND NITRO-CELLULOSE IN DIFFERENT PROPORTIONS.

Composition of Explosives.		Calories per cent.
Nitro-cellulose ($N = 13.3$ per cent.).	Nitro-glycerine.	
100 per cent. dry pulp	0	1061
100 " gelatinised	0	922
90 "	10 per cent.	1044
80 "	20 "	1159
70 "	30 "	1267
60 "	40 "	1347
50 "	50 "	1410
40 "	60 "	1467
0 "	100 "	1652
Nitro-cellulose ($N=12.24$ per cent.)	Nitro-glycerine.	
80 per cent.	20 per cent.	1062
60 "	40 "	1288
50 "	50 "	1349
40 "	60 "	1405

Nitro-cellulose (N = 13.3 per cent.)	Nitro-glycerine	Vaseline
55 per cent.	40 per cent.	5 per cent. 1134
35 "	60 "	5 " 1280

TABLE OF RESULTS OBTAINED BY LIEUT. W. WALKE., OF THE ARTILLERY, U.S.A, WITH QUINAN'S PRESSURE GAUGE.

Nitro-glycerine being taken as 100. (From *U.S. Naval Inst. Jour.*)

Name of Explosive.	Compression of Lead Inch.	Order of Strength.	
Explosive gelatine	0.585	106.17	
Hellhoffite	0.585	106.17	
Nitro-glycerine	0.551	100.00	Standard, N.G.
Nobel's smokeless powder	0.509	92.38	
Nitro-glycerine	0.509	92.37	
Gun-cotton	0.458	83.12	U.S. naval torpedo gun-cotton
Gun-cotton	0.458	83.12	Stowmarket.
Nitro-glycerine	0.451	81.85	Vouges, N.G.
Gun-cotton	0.448	81.31	
Dynamite No. 1	0.448	81.31	
Dynamite de Traul	0.437	79.31	
Emmensite	0.429	77.86	
Amide powder	0.385	69.87	
Oxonite	0.383	69.51	
Tonite	0.376	68.24	G.C. 52.5%, and

			Ba(NO$_{3}$)$_{2}$, 47.5%
Bellite	0.362	65.70	
Rack-a-rock	0.340	61.71	
Atlas powder	0.333	60.43	
Ammonia dynamite	0.332	60.25	
Volney's powder No. 1	0.322	58.44	Nitrated naphthalene.
" No. 2	0.294	53.18	" "
Melinite	0.280	50.82	Picric acid 70%, and
			sol. nitro-cotton 30%.
Silver fulminate	0.277	50.27	
Mercury	0.275	49.91	
Mortar powder	0.155	28.13	

~Composition of some of the Explosives in Common Use.~

~Ordinary Dynamite.~

Nitro-Glycerine 75 per cent.
Kieselguhr 25 "

~Amvis.~

Nitrate of Ammonia 90 per cent.
Chloro-di-nitro Benzene 5 "
Wood Pulp 5 "

~Ammonia Nitrate Powder.~

Nitrate of Ammonia 80 per cent.
Chlorate of Potash 5 "
Nitro-Glucose 10 "
Coal Tar 5 "

~Celtite.~

Nitro-Glycerine 56-59 parts.
Nitro-Cotton 2-3.5 "
KNO$_{3}$ 17-21 "

Wood Meal 8-9 "
Ammonium Oxalate 11-13 "
Moisture 0.5-1.5 "

~Atlas Powders.~

Sodium Nitrate 2.0 per cent.
Nitro-Glycerine 75.0 "
Wood Pulp 21.0 "
Magnesium Carbonate 2.0 "

~Dauline.~

Nitro-Glycerine 50 per cent.
Sawdust 30 "
Nitrate of Potash 20 "

~Vulcan Powder.~

Nitro-Glycerine 30 per cent.
Nitrate of Soda 52.5 "
Sulphur 7.0 "
Charcoal 10.5 "

~Vigorite.~

Nitro-Glycerine 30 per cent.
Nitrate of Soda 60 "
Charcoal 5 "
Sawdust 5 "

~Rendrock.~

Nitrate of Potash 40 per cent.
Nitro-Glycerine 40 "
Wood Pulp 13 "
Paraffin or Pitch 7 "

~Ammonia Nitrate Powder.~

Ammonia Nitrate 80 per cent.
Potassium Chlorate 5 "
Nitro-Glucose 10 "
Coal Tar 5 "

~Hercules Powders.~

Nitro-Glycerine 75 to 40 per cent.
Sugar 1 " 15.66 "
Chlorate of Potash 1.05 " 3.34 "
Nitrate of Potash 2.10 " 31.00 "
Carbonate of Magnesia 20.85 " 10.00 "

~Carbo-Dynamite.~

Nitro-Glycerine 90 per cent.
Charcoal 10 "

~Geloxite (Permitted List).~

Nitro-Glycerine 64-54 parts.
Nitro-Cotton 5-4 "
Nitrate of Potash 22-13 "
Ammonium Oxalate 15-12 "
Red Ochre 1-0 "
Wood Meal 7-4 "

The Wood Meal to contain not more than 15% and not less than 5% moisture.

~Giant Powder.~

Nitro-Glycerine 40 per cent.
Sodium Nitrate 40 "
Rosin 6 "
Sulphur 6 "
Guhr 8 "

~Dynamite de Trauzel.~

Nitro-Glycerine 75 parts.
Gun-Cotton 25 "
Charcoal 2 "

~Rhenish Dynamite.~

Solution of N.G. in Naphthalene 75 per cent.
Chalk, or Barium Sulphate 2 "
Kieselguhr 23 "

~Ammonia Dynamite.~

Ammonia Nitrate 75 parts.
Paraffin 4 "
Charcoal 3 "
Nitro-Glycerine 18 "

~Blasting Gelatine.~

Nitro-Glycerine 93 per cent.
Nitro-Cotton 3 to 7 "

~Gelatine Dynamite.~

Nitro-Glycerine 71 per cent.
Nitro-Cotton 6 "
Wood Pulp 5 "
Potassium Nitrate 18 "

~Gelignite.~

Nitro-Glycerine 60 to 61 per cent.
Nitro-Cotton 4 " 5 "
Wood Pulp 9 " 7 "
Potassium Nitrate 27 "

~Forcite.~

Nitro-Glycerine 49 per cent.
Nitro-Cotton 1.0 "

Sulphur 1.5 "
Tar 10.0 "
Sodium Nitrate 38.0 "
Wood Pulp 5 "
 (The N.-G., &c., varies.)

~Tonite No. 1.~

Gun-Cotton 52-50 per cent.
Barium Nitrate 47-40 "

~Tonite No. 2.~

Contains Charcoal also.

~Tonite No. 3.~

Gun-Cotton 18 to 20 per cent.
$Ba(NO_3)_2$ 70 " 67 "
Di-nitro-Benzol 11 " 13 "
Moisture 0.5 " 1 "

~Carbonite.~

Nitro-Glycerine 17.76 per cent.
Nitro-Benzene 1.70 "
Soda 0.42 "
KNO_3 34.22 "
$Ba(NO_3)_2$ 9.71 "
Cellulose 1.55 "
Cane Sugar 34.27 "
Moisture 0.36 "

———

99.99

~Roburite.~

Ammonium Nitrate 86 per cent.
Chloro-di-nitro-Benzol 14 "

~Faversham Powder.~

Ammonium Nitrate 85 per cent.
Di-nitro-Benzol 10 "
Trench's Flame-extinguishing Compound 5 "

~Favierite No. 1.~

Ammonium Nitrate 88 per cent.
Di-nitro-Naphthalene 12 "

~Favierite No. 2.~

No. 1 Powder 90 per cent.
Ammon. Chloride 10 "

~Bellite.~

Ammonium Nitrate 5 parts.
Meta-di-nitro-Benzol 1 "

~Petrofacteur.~

Nitro-Benzene 10 per cent.
Chlorate of Potash 67 "
Nitrate of Potash 20 "
Sulphide of Antimony 3 "

~Securite.~

Mixtures of Meta-di-nitro-Benzol 26 per cent. and Nitrate of Ammonia 74 "

~Rack-a-Rock.~

Potassium Chlorate 79 parts.
Mono-nitro-Benzene 21 "

~Oxonite.~

Nitric Acid (sp. gr. 1.5) 54 parts.
Picric Acid 46 "

~Emmensite.~

Emmens Acid 5 parts.
Ammonium Nitrate 5 "
Picric Acid 6 "

~Brugère Powder.~

Ammonium Picrate 54 per cent.
Nitrate of Potash 46 "

~Designolle's Torpedo Powders.~

Potassium Picrate 55 to 50 per cent.
Nitrate of Potash 45 " 50 "

~Stowite.~

Nitro-Glycerine 58 to 61 parts.
Nitro-Cotton 4.5 " 5 "
Potassium Nitrate 18 " 20 "
Wood Meal 6 " 7 "
Oxalate of Ammonia 11 " 15 "

The Wood Meal shall contain not more than 15% and not less than 5% by weight of moisture. The explosive shall be used only when contained in a non-water-proofed wrapper of parchment—No. 6 detonator.

~Faversham Powder.~

Nitrate of Ammonium 93 to 87
Tri-nitro-Toluol 11 " 9
Moisture 1 " —

~Kynite.~

Nitro-Glycerine 24-26 parts.
Wood-Pulp 2.5-3.5 "
Starch 32.5-3.5 "
Barium Nitrate 31.5-34.5 "
$CaCO_{3}$ 0-0.5 "
Moisture 3.0-6.0 "

Must be put up only in water-proof parchment paper, and No. 6 electric detonator used.

~Rexite.~

Nitro-Glycerine 6.5-8.5 parts.
Ammonium Nitrate 64-68 "
Sodium Nitrate 13-16 "
Tri-nitro-Tolulene 6.5-8.5 "
Wood Meal 3-5 "
Moisture .5-1.4 "

Must be contained in water-proof case (stout paper), water-proofed with
Resin and Cerasin—No. 6 detonator.

~Withnell Powder.~

Ammonium Nitrate 88-92 parts.
Tri-nitro-Toluene 4-6 "
Flour (dried at 100° C.) 4-6 "
Moisture 0-15 "

Only to be used when contained in a linen paper cartridge, water-proofed with Carnuba Wax, Parrafin—No. 7 detonator used.

~Phenix Powder.~

Nitro-Glycerine 28-31 parts.
Nitro-Cotton 0-1 "
Potassium Nitrate 30-34 "
Wood Meal 33-37 "

Moisture 2-6 "

~SMOKELESS POWDERS.~

~Cordite.~

Nitro-Glycerine 58 per cent. +or- .75
Nitro-Cotton 37 " +or- .65
Vaseline 5 " +or- .25

~Cordite, M.D.~

Nitro-Glycerine 30 per cent. +or- 1
Nitro-Cotton 65 " +or- 1
Vaseline 5 " +or- .25

Analysis of—
  By W. Mancab and A.E. Leighton.

~E.C. Powder.~

Nitro-Cotton 79.0 per cent.
Potassium Nitrate 4.5 "
Barium Nitrate 7.5 "
Camphor 4.1 "
Wood Meal 3.8 "
Volatile Matter 1.1 "

~Walarode Powder.~

Nitro-Cotton 98.6 per cent.
Volatile Matter 1.4 "

~Kynoch's Smokeless.~

Nitro-Cotton 52.1 per cent.
Di-nitro-Toluene 19.5 "
Potassium Nitrate 1.4 "
Barium Nitrate 22.2 "
Wood Meal 2.7 "

Ash 0.9 "
Volatile Matter 1.2 "

~Schultze.~

Nitro-Lingin 62.1 per cent.
Potassium Nitrate 1.8 "
Barium Nitrate 26.1 "
Vaseline 4.9 "
Starch 3.5 "
Volatile Matter 1.0 "

~Imperial Schultze.~

Nitro-Lignin 80.1 per cent.
Barium Nitrate 10.2 "
Vaseline 7.9 "
Volatile Matter 1.8 "

~Cannonite.~

Nitro-Cotton 86.4 per cent.
Barium Nitrate 5.7 "
Vaseline 2.9 "
Lamp Black 1.3 "
Potassium Ferro-cyanide 2.4 "
Volatile Matter 1.3 "

~Amberite.~

Nitro-Cotton 71.0 per cent.
Potassium Nitrate 1.3 "
Barium Nitrate 18.6 "
Wood Meal 1.4 "
Vaseline 5.8 "

~Sporting Ballistite.~

Nitro-Glycerine 37.6 per cent
Nitro-Cotton 62.3 "

Volatile Matter 0.1 "

The following is a complete List of the Permitted Explosives as Defined in the Schedules to the Explosives in Coal Mines Orders of the 20th December 1902, of the 24th December 1903, of the 5th September 1903, and 10th December 1903:—

Albionite.
Ammonal.
Ammonite.
Amvis.
Aphosite.
Arkite.
Bellite No. 1.
Bellite No. 2.
Bobbinite.
Britonite.
Cambrite.
Carbonite.
Clydite.
Coronite.
Dahmenite A.
Dragonite.
Electronite.
Faversham Powder.
Fracturite.
Geloxite.
Haylite No. 1.
Kynite.
Negro Powder.
Nobel's Ardeer Powder.
Nobel Carbonite.
Normanite.
Pit-ite.
Roburite No. 3.
Saxonite.
Stow-ite.
Thunderite.
Victorite.

Virite.
West Falite No. 1.
West Falite No. 2.